SpringerBriefs in Materials

Series Editors

Sujata K. Bhatia, University of Delaware, Newark, DE, USA

Alain Diebold, Schenectady, NY, USA

Juejun Hu, Deparment of Materials Science and Engineering, Massachusetts Institute of Technology, Cambridge, MA, USA

Kannan M. Krishnan, University of Washington, Seattle, WA, USA

Dario Narducci, Department of Materials Science, University of Milano Bicocca, Milano, Italy

Suprakas Sinha Ray ⓘ, Centre for Nanostructures Materials, Council for Scientific and Industrial Research, Brummeria, Pretoria, South Africa

Gerhard Wilde, Altenberge, Nordrhein-Westfalen, Germany

The SpringerBriefs Series in Materials presents highly relevant, concise monographs on a wide range of topics covering fundamental advances and new applications in the field. Areas of interest include topical information on innovative, structural and functional materials and composites as well as fundamental principles, physical properties, materials theory and design. SpringerBriefs present succinct summaries of cutting-edge research and practical applications across a wide spectrum of fields. Featuring compact volumes of 50 to 125 pages, the series covers a range of content from professional to academic. Typical topics might include

- A timely report of state-of-the art analytical techniques
- A bridge between new research results, as published in journal articles, and a contextual literature review
- A snapshot of a hot or emerging topic
- An in-depth case study or clinical example
- A presentation of core concepts that students must understand in order to make independent contributions

Briefs are characterized by fast, global electronic dissemination, standard publishing contracts, standardized manuscript preparation and formatting guidelines, and expedited production schedules.

More information about this series at http://www.springer.com/series/10111

Lan'er Wu · Fenglan Han · Guiqun Liu

Comprehensive Utilization of Magnesium Slag by Pidgeon Process

Lan'er Wu
School of Materials Science
and Engineering
North Minzu University
Yinchuan, Ningxia, China

Fenglan Han
School of Materials Science
and Engineering
North Minzu University
Yinchuan, Ningxia, China

Guiqun Liu
School of Materials Science
and Engineering
North Minzu University
Yinchuan, Ningxia, China

ISSN 2192-1091   ISSN 2192-1105 (electronic)
SpringerBriefs in Materials
ISBN 978-981-16-2173-4   ISBN 978-981-16-2171-0 (eBook)
https://doi.org/10.1007/978-981-16-2171-0

This Springer imprint is published by the registered company Springer Nature Singapore Pte Ltd.
The registered company address is: 152 Beach Road, #21-01/04 Gateway East, Singapore 189721, Singapore

Preface

Magnesium is a promising material which has light weight, high specific strength, good ductility, good thermal conductivity and so on. China contributes more than 80% of the global magnesium production to the world each year. Most of Chinese magnesium metallurgy enterprises employ silicothermic reduction process—Pidgeon process, which produces huge amount of the reduction slag. The magnesium slag is not only the industrial waste polluting environment, but also the valuable raw materials which can be re-used. With rapid economic development in China resent years, there are increasingly urgent needs for comprehensive utilization of large amount of industrial solid waste. It has become the bottleneck that affecting sustained development of local economy. The team of authors in North Minzu University focus on the research of local industrial solid waste such as magnesium slag, electrolytic manganese residue and fly ash in recent years. The current book introduces the research results of the author team on recycling and utilization of magnesium slag. The main experimental data and cases come from the findings of the several national research projects such as National Program on Key Basic Research Project of China, International S&T Cooperation Project of China National 12th "Five-Year" Technology Support Programs undertaken by the authors. This book provides detail description of the research projects including technical roadmap, experimental data, and industrial test conditions and results and so on. It is expected that the book could provide a reference to scientists and engineers in relevant field and readers with interest.

There are five chapters in the book. Chapter 1 mainly introduces magnesium resources of the world and enterprise layout, the history of magnesium industry. This chapter is written by Dr. Guiqun Liu. Chapter 2 introduced the magnesium metallurgical technology. There are two main magnesium metallurgical methods: electrolytic method and reduction method. As the mainstream technology widely used in Chinese magnesium metallurgy industry, Pidgeon process are described. Chapter 2 is also written by Dr. Guiqun Liu. Chapter 3 is written by Prof. Lan'er Wu. The generation process and behavior of magnesium slag via Pidgeon process are described in this chapter. Compositions, phases, particle distribution and harmful elements in magnesium slag were analyzed. The state and development of the techniques of treatment and effective use of magnesium slag were reviewed. Chapter 4 relates formation mechanism of fine magnesium slag dust, chemical stabilization of the slag and fluoride-free mineralizers for magnesium smelting. Chapter 5 introduces comprehensive utilization attempts of magnesium slag made by the team of the authors. Chapters 4 and 5 are written by Prof. Lan'er Wu and Prof. Fenglan Han.

Postgraduate student Chen Hao has help on some of text editing and diagrams drawing.

The authors express their thanks to the financial support of Ministry of Science and Technology of China and Science and Technology Department of Ningxia Autonomous Region.

Yinchuan, China   Lan'er Wu
                                                                       Fenglan Han
                                                                       Guiqun Liu

Brief Introduction

Magnesium metal is widely used in aviation, aerospace, transportation and electronics because of its excellent properties, such as light weight, high specific strength, good ductility, good thermal conductivity and so on. In recent years, China become the largest magnesium raw material production country in the world. Its annual output of Magnesium reaches 80% of that of the whole world. Most of Chinese magnesium enterprises employ silicothermic reduction process—Pidgeon process, which produces huge amount of the reduction slag. The magnesium slag is not only the industrial waste polluting environment, but also the useful raw materials which can be recycled. In this book, magnesium resources of the world and enterprise layout are introduced. The process of magnesium silicothermic reduction and how the magnesium slag produced are described. As main point, harmless recycling of magnesium slag is investigated in details. Many examples and experimental data in this book came from the author team's research programs. It is expected that the book could provide precious reference to scientists and engineers in the field of recycling and environmentally friendly use of the industrial solid wastes. In addition, it will also give information to the students who is interested in relevant field.

Contents

Chapter 1
Overview of Magnesium Metallurgy

Abstract Human beings discovered magnesium compounds as early as the seventeenth century. In later years, the magnesium alloy has been used in our daily life. In this section, we discuss the magnesium mineral resource, magnesium smelting technology, the history of magnesium industry and the comprehensive utilization of magnesium slag. We believe that it is worthwhile to investigate the magnesium and ralated technology which will make our life better.

Keywords Magnesium · Magnesium mineral · Magnesium industry · Magnesium smelting · Magnesium slag

1.1 Introduction

1.1.1 Properties and Main Uses of Magnesium Metal

Human beings discovered magnesium compounds as early as the seventeenth century. Antoine Lavosier, a French scientist, theoretically inferred that an ore with unknown composition (ore containing alumina and magnesium oxide) contained a new metal element, but the magnesium could not be extracted using the reducing agents known at that time because of the strong bond between magnesium and oxygen atoms.

Magnesium is a kind of light metal with high chemical activity. There is a huge reserve of magnesium ores widely distributed across the globe. Table 1.1 shows the distribution of elements in the earth's crust. Table 1.2 shows the main physical properties of magnesium. Because it has the advantages of light weight, high specific strength, good ductility, good damping and machinability, a strong electromagnetic shielding effect, good shock absorption, good thermal conductivity and thermal fatigue performance, and is easy to recycle, magnesium is widely used in aviation, aerospace, transportation, electronics, and other fields. The magnesium alloy is far superior to the aluminum alloy in electromagnetic shielding and shock absorption. Compared with fiber-reinforced plastics, the magnesium alloy has lower specific strength but higher specific stiffness. Because of its high chemical activity, magnesium can also be used as a reducing agent in the production processes of refractory metals (Ti, Zr, Be, U, and HF). Because magnesium has an extremely

L. Wu et al., *Comprehensive Utilization of Magnesium Slag by Pidgeon Process*,
SpringerBriefs in Materials, https://doi.org/10.1007/978-981-16-2171-0_1

Table 1.1 Distribution of chemical elements in the earth's crust

Element	Mass content/%	Element	Mass content/%
Oxygen	48.06	Calcium	3.45
Silicon	26.30	Sodium	2.74
Aluminum	7.73	Potassium	2.47
Iron	4.75	Magnesium	2.00
Hydrogen	0.76	Other	0.76

Table 1.2 Main physical properties of magnesium

Melting point/°C	648
Boiling point/°C	1107
Relative density (water = 1)	1.74
Appearance and character	Silver white metallic powder
Solubility	It is insoluble in water and alkali solution and soluble in acid

high affinity with sulfur, it can also be used as a desulfurizer. Magnesium performs unique deoxidation and purification functions in the production of alloy materials that involve Cu, Ni, Zn, and rare earth elements. In addition to being used as a reducing agent of refractory metals and an additive of alloys, magnesium can also be used as a nodularizing agent of nodular cast iron, neutralizer of lubricating oil, and large-capacity energy storage materials. Moreover, magnesium also plays a role in automobiles, electronic communication, aerospace, and other fields thanks to its light weight and ease of processing (cutting and die casting) [1].

Because of the above excellent properties, metal magnesium is touted as "the most promising lightweight engineering metal material in the twenty-first century" [2]. Magnesium and its alloys were first used in the aviation industry during the First World War. Despite its long history of commercial application, the development of magnesium alloys has been slow compared with that of aluminum alloys. Since 2000, the reserves of many traditional metals have been drying up. As governments across the world adopt development strategies based on energy conservation and environmental protection, the superb performance and economic benefits of magnesium and its alloys have attracted extensive attention. Many countries have invested a lot of human and financial resources in the research and development of magnesium-based materials, and the research results have been widely applied in various industrial fields. This is mainly reflected in the following areas: (1) The automobile industry: In response to the increasingly stringent requirements in almost all countries regarding exhaust emission, fuel consumption, and noise, automobile manufacturers seek to replace common components made of steel and lead alloys with ones made of magnesium alloys. As magnesium alloys are 77% lighter than steel and 36% lighter than lead alloys, the use of magnesium alloys can significantly reduce vehicle weight, thus reducing fuel consumption and exhaust emission. At present, North America ranks no. 1 in terms of magnesium alloy consumption in the automobile industry, with an annual growth rate of 30%. In China, Shanghai Automotive Company takes the lead in applying magnesium alloys in the production of automobile transmission cases, bringing the annual magnesium consumption to a level higher than 2000 tonnes. (2) Electronic products: Nowadays, electronic products have become necessities in daily life, and electronic products are evolving towards small size and low cost. Compared with traditional engineering plastics that are widely used in electronic products, magnesium alloys have unique advantages in the miniaturization of electronic products thanks to a series of merits, including ease of being made into high-performance thin walls, high strength, and strong impact resistance. Magnesium alloys have been used in producing cases and parts of electronic products, and the market has been growing steadily. (3) Aerospace industry: Because of its light weight, magnesium alloys were used in the aviation industry during the First World War to reduce the weight of aircraft. Today, magnesium alloys are still used in the manufacture of some parts of military and civil aircraft, such as the support structure, which helps improve the dynamic performance of aircraft and reduce the weight. The application scope of magnesium alloys in the aerospace field will be expanded as the properties of magnesium alloys are gradually improved. (4) Other fields: Magnesium alloys have excellent mechanical properties and good formability, and magnesium is one of the essential metal elements of the human body. Thus, magnesium alloys can be used as a medical implant material. Because magnesium is light and comfortable to touch, it is also suitable for producing bicycles, wheelchairs, and other appliances that are used in daily life.

At present, China is the largest producer and exporter of magnesium in the world, accounting for more than 85% of the world's magnesium output. China's magnesium production capacity is highly concentrated in regions with low energy costs, such as Shaanxi, Shanxi, Xinjiang, and Inner Mongolia. As the regulations related to environmental protection become more and more stringent and competition becomes increasingly fierce, some magnesium smelting firms exit the market every year, bringing down the total number of magnesium smelting firms in China. At the end of 2018, there were over 80 magnesium smelting firms in China, most of which have small production capacity. The top ten firms accounted for only 37.7% of the total output, indicating a very low level of concentration in the industry.

According to statistics, China's output of metallic magnesium in 2017 was 912,600 tonnes, and the output in 2018 was 863,000 tonnes, a decrease of 5.44% compared with the same period of the previous year. Figure 1.1 shows the variation of China's magnesium output from 2014 to 2018.

According to customs statistics, China's export volume of metal magnesium in 2018 was 409,800 tonnes, which is a year-on-year decrease of 9.78%; the export amount of metal magnesium in the same year was 1.064 billion US dollars, which is a year-on-year decrease of 2.51%. In 2018, China's import of metal magnesium was 462.35 tonnes, which is a year-on-year increase of 27.83%; the import amount of magnesium in the same year was 12.4201 million US dollars, which is an increase of 44.18% year-on-year. Table 1.3 shows the statistical figures about China's import and export of magnesium from 2010 to 2018.

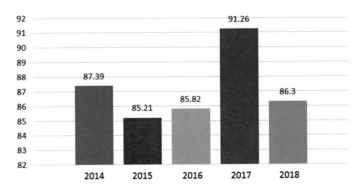

Fig. 1.1 Variations in China's magnesium output from 2014 to 2018 (in 10,000 tonnes)

Table 1.3 Statistical figures about China's import and export of magnesium from 2010 to 2018

Year	Import amount/USD	Import volume/kg	Export amount/USD	Export volume/kg
2010	$5,042,727	1,021,566	$1,062,226,022	383,980,095
2011	$6,198,693	1,182,031	$1,239,150,801	400,076,252
2012	$4,792,884	694,501	$1,157,021,413	371,084,410
2013	$3,721,902	396,831	$1,187,366,008	411,122,706
2014	$9,387,038	1,325,233	$1,171,995,186	434,996,136
2015	$9,535,648	1,623,543	$1,006,675,498	405,435,198
2016	$11,433,999	736,283	$852,234,148	356,536,970
2017	$8,614,564	361,683	$1,058,038,865	454,191,109
2018	$12,420,097	462,348	$1,031,497,935	409,788,162

It is estimated that the total demand for magnesium in China's metallurgical industry will be 278,200 tonnes in 2018, the demand in the metal processing fields (castings, die castings, and profiles) will be 156,400 tonnes, and the demand in other fields will be 12,000 tonnes. According to the data in recent years, the demand for metal magnesium has been increasing year by year, and the domestic supply and demand have reached a balance. Figure 1.2 shows the actual consumption and supply-demand balance of magnesium in China from 2014 to 2018.

1.2 Magnesium Mineral Resources

1.2.1 Global Distribution of Magnesium Mineral Resources

Magnesium resources are abundant and widely distributed in the world. Magnesium is one of the most abundant light metal elements on earth. Among all of the natural

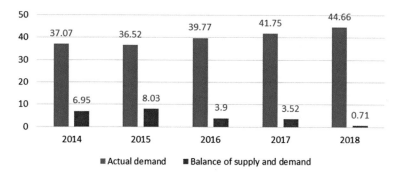

Fig. 1.2 Actual consumption and supply-demand balance of magnesium in China from 2014 to 2018 (in 10,000 tonnes)

elements on earth, magnesium ranks no. 8 in terms of content. The content of magnesium in the earth's crust is 2%, and by content, magnesium in seawater ranks no.3. Although magnesium is rich in nature and it is one of the most widely distributed elements in the earth's crust (ranking no. 8), it can only exist in nature in the form of compounds because of its high chemical activity. There are more than 200 kinds of ores that contain magnesium compounds, but only a few can be used as raw materials for magnesium smelting.

Magnesium is widely distributed across the world, and magnesium salt resources are extremely rich worldwide; they are available mainly in the forms of solid mineral resources and liquid mineral resources. Solid mineral resources mainly include magnesite, dolomite, serpentine, talc, brucite, and a small amount of other sedimentary minerals. Liquid mineral resources mainly exist in sea water, natural salt lake water, brine, etc., that occupy a large portion of the earth's surface. Liquid mineral resources are virtually inexhaustible. Natural brine can be regarded as a kind of recyclable resource, and so, the amount of magnesium minerals mined by humans will be regenerated in a relatively short time. Although more than 60 minerals contain magnesium, the majority of magnesium used in the world comes from dolomite, magnesite, brucite, carnallite, and olivine, followed by seawater bittern, salt lake brine, and underground brine. The current reserves of magnesium minerals can fully meet the human demand for magnesium, and there will not be shortage of supply in the foreseeable future. Table 1.4 shows the distribution of various types of magnesium mineral resources in the world [3].

Among the magnesium resources, magnesite is the principle magnesium mineral with value in industrial applications. According to the data released by the United States Geological Survey (USGS) in 2015, the world's proven reserves of magnesite minerals amount to 12 billion tonnes, among which 2.4 billion tonnes are exploitable reserves. Countries with rich reserves of magnesium minerals include Russia (650 million tonnes, accounting for 27%), China (500 million tonnes, accounting for 21%), and South Korea (450 million tonnes, accounting for 19%). The largest and highest-quality magnesite deposit in the world is located in Dashiqiao, Liaoning Province, China. Table 1.5 shows magnesite reserves of countries.

Table 1.4 Distribution of magnesium mineral resources

Mineral	Molecular formula	Magnesium content	Countries with the largest reserves
Magnesium silicate			
Serpentine	$3MgO \cdot 2SiO_2 \cdot 2H_2O$	26.3	Former Soviet Union, Canada
Olivine	$(MgFe)_2 \cdot SiO_4$	34.6	Italy, Norway
Talc	$3MgO \cdot 4SiO_2 \cdot 2H_2O$	19.2	USA, Spain
Magnesium carbonate			
Magnesite	$MgCO_3$	28.8	China, India, USA
Dolomite	$MgCO_3 \cdot CaCO_3$	13.2	China, the former Soviet Union, the United States
Magnesium chloride			
Bischofite	$MgCl_2 \cdot 6H_2O$	12.0	China, the former Soviet Union, the United States
Carnallite	$MgCl_2 \cdot KCl \cdot 6H_2O$	8.8	
Magnesium sulphate			
Magnesium sulfate	$MgSO_4 \cdot H_2O$	17.6	
Jarosite	$MgSO_4 \cdot KCl \cdot 6H_2O$	9.8	The former Soviet Union, China, the United States
Polyhalite	$MgSO_4 \cdot K_2SO_4 \cdot 2CaSO_4 \cdot 2H_2O$	4.0	Germany
Anhydrous jarosite	$2MgSO_4 \cdot K_2SO_4$	11.7	
Albite	$MgSO_4 \cdot NaSO_4 \cdot 4H_2O$	7.0	China

1.2.2 Distribution of Magnesium Mineral Resources in China

China is one of the countries with the most abundant magnesium resources in the world, with total reserves accounting for 22.5% of the world's total. China has many varieties of magnesium mineral resources, including magnesite, dolomite, and salt lake magnesium salt, and they are widely distributed across the country. The magnesium minerals that have been mined and used include dolomite, magnesite, carnallite, and potassium magnesium salts in Qarhan, Qinghai Province. There are 27 mining areas with proven magnesite reserves in China, and these are distributed in 9 provinces (regions). Liaoning Province ranks no. 1 in terms of magnesite reserves, accounting for 85.6% of the country's total. Liaoning, Shandong, Tibet, Xinjiang, and Gansu, also have large magnesite reserves.

At present, the proven dolomite reserves in China amount to 4 billion tonnes, the magnesite reserves are about 3.47 billion tonnes, and the prospective reserves of magnesium salt resources in salt lake areas are more than 8 billion tonnes (the

Table 1.5 Magnesite reserves of countries in the world (100 million tonnes)

Country	Reserve
U.S.A	0.1
Australia	0.95
Austria	0.15
Brazil	0.86
China	5
Greece	0.8
India	0.2
Korea	4.5
Russia	6.5
Slovakia	0.35
Spain	0.1
Turkey	0.49
Other countries	3.9
Global total	23.9

magnesium salt reserve in Qarhan Salt Lake in Qinghai Province is about 4.82 billion tonnes) [3]. The total magnesium resources of China account for 22.5% of the world's total, ranking no.1 in the world [4–6].

1. **Dolomite**

China is also rich in magnesium-bearing dolomite, with proven reserves of more than 4 billion tonnes. Dolomite resources are found in all provinces and regions of China, especially Shanxi, Ningxia, Henan, Jilin, Qinghai, and Guizhou. Dolomite deposits can be divided into hydrothermal type and sedimentary type according to their properties. Hydrothermal deposits have been widely developed in eastern Liaoning region and Jiaodong region, and sedimentary deposits are mainly distributed in Shanxi, Henan, Hunan, Hubei, Guangxi, Guizhou, Ningxia, Jilin, Qinghai, Yunnan, and Sichuan provinces [7].

Most dolomites are secondary sediments that resulted from the metasomatic process between a magnesium-containing solution and limestone. Only in high salinity lakes can thick dolomites and primary sedimentary dolomites be formed directly. Dolomite is a compound salt mineral composed of magnesium carbonate and calcium carbonate. The theoretical mass fraction is 21.7% dolomite, 30.4% $CaCO_3$, and 47.9% CO_2 [8], and the mass ratio of CaO to MgO is 1.394. In natural dolomite minerals, the mass ratio of CaO to MgO is in the range of 1.4–1.7, the relative density is 2.8–2.9 g $\times$ cm^{-3}, and the Mohs hardness is 3.4–4. Dolomite crystals are hexagonal, their common color is white with a tint of yellow or brown, and they have a glass luster.

Because dolomites contain impurities, their chemical and physical properties may vary. The structures of dolomites can be roughly divided into two types [9, 10]: one is a hexagonal rhombic structure, and the other is an amorphous network structure. Compared with dolomites that have a network structure, the calcined dolomites with a hexagonal rhombic structure are easy to grind, do not stick to the grinder, have low reaction activity [11], and are brittle. The dolomites with an amorphous network structure retain the structural characteristics of dolomite after calcination, have low lattice energy, and have lower heat absorption than dolomites with hexagonal rhombic structure during thermal decomposition. Because of these properties, the calcination time of dolomites with an amorphous network structure is much shorter than that of dolomites with a hexagonal rhombic structure.

Dolomites are widely distributed in the world. In addition to China, the major producers of dolomites are Switzerland, Italy, North England, and Mexico. When dolomite is used as a raw material for magnesium smelting, the thermal reduction method is usually used. Dolomites can also be used in building materials, ceramics, the chemical industry, and other fields.

China is also rich in magnesium-bearing dolomites, with more than 4 billion tonnes of proven reserves and 208 explored mining areas. The dolomite resources are widely distributed across the country. Almost all provinces and regions in China have dolomite mines, represented by Hunan, Sichuan, Shandong, Hebei, Shanxi, Liaoning, Jilin, and Inner Mongolia. At present, most of the deposits have been exploited. According to the Overall Plan of Mineral Resource Exploitation in Shanxi Province (2016–2020), the retained reserves of magnesium minerals (magnesium dolomite) in Shanxi Province were 845 million tonnes by the end of 2015, accounting for 30% of the country's total and ranking no. 1 in China. In recent years, Shannxi Province has developed a circular industrial chain of using the exhaust gas of a blue coal (a kind of clean coal produced using high-quality coals mined from the mines in Fugu, Shenmu, and other places in Shaanxi Province) production facility to produce ferrosilicon and then using ferrosilicon to reduce metal magnesium, achieving great cost savings. Consequently, the raw magnesium output of Shaanxi accounted for 62% of China's total, and the share of Shanxi Province declined to 14%.

Because dolomite can be used as a refractory material, electrical insulation material, chemical building material, advanced ceramic material, and sealing material, it has the potential of being used widely in various fields. China has huge reserves of high-quality dolomite with a potentially wide application scope, and the magnesium alloy products that are developed with dolomite as a raw material play an important role in China's economic and social development.

The composition of dolomite varies depending on place of origin. Table 1.6 shows the compositions of dolomite ores mined from several typical dolomite mining areas.

At present, all of the dolomite mines are open-pit mining, and most of them supply building materials for residents. The Baiyunyan dolomite deposit in Nanjing, Jiangsu Province is the first deposit to have been exploited, and it has developed into a large-scale mine with a high level of mechanization. The dolomite ores are high quality, and more than 50% of the ores have a MgO content higher than 20% and a SiO_2 content less than 2%. With an annual mining capacity of over one million tonnes, the mine mainly supplies dolomite to serve as a refractory material and flux in the

Table 1.6 Compositions of dolomite ores mined from several typical dolomite mining areas in China

Place of origin	SiO_2	Al_2O_3	Fe_2O_3	CaO	MgO	Loss on ignition
Great Stone Bridge	0.32	0.39	0.89	30.28	21.72	47.08
Wulong Spring	0.03	0.05	0.34	31.75	20.02	47.10
Lacao Mountain, Guyang	1.53	0.14	0.75	30.10	19.48	46.13
Yutian	0.27	0.10	0.03	30.52	21.91	46.80
Zhoukoudian	0.55	0.08	0.16	29.06	21.93	46.24
Zhen'an, Shaanxi Province	1.90	0.58	0.58	30.00	21.00	46.94
Zhenjiang	1.17	0.37	0.18	30.80	21.16	47.07

production processes of Baoshan Iron and Steel Company and Ma'anshan Iron and Steel Company.

In the late 1980s, dolomite was used as a raw material for magnesium smelting in silicothermic magnesium smelting plants, and magnesium smelting plants usually use magnesium ores mined from nearby mines. To date, there is still no unified quality standard for dolomite that is used in magnesium smelting. Before a newly-built magnesium smelting plant is put into operation, it is necessary to carry out lab-testing on the quality of dolomite ore to be used in the magnesium smelting process because the ore quality has a great impact on the technical and economic indicators of magnesium production. First, the chemical composition of the ore should meet the requirement of MgO > 20%, $Fe_2O_3 + Al_2O_3 \leq 1\%$, $SiO_2 \leq 1\%$, and $Na_2O + K_2O \leq 0.1\%$. Second, the structural characteristics of the ore should be considered. The ore structure has a certain influence on the magnesium smelting processes, such as calcination and ball making.

With the rapid development of silicothermic magnesium smelting plants in recent years, the consumption of dolomite mineral resources also increased. In 1992, the total consumption in China was about 100,000–130,000 tonnes, and this figure reached 650,000–7,150,000 tonnes in 2007, which is a 65-fold increase.

2. Magnesite mineral

According to the data released by USGS in 2015, the global magnesite output in 2014 was 6.97 million tonnes, which is an increase of 60,000 tonnes over the same period of the previous year. China is the biggest producer of magnesite, accounting for 70.3% of the world's total. China has a large magnesite reserve, second only to that of Russia. Most of China's magnesite reserves are concentrated in large deposits in a few regions. There are 27 mining areas with proven magnesite reserves in China, and these are distributed in 9 provinces (regions). Liaoning Province ranks No. 1 in terms of magnesite reserves, accounting for 85.6% of the country's total. In addition to Liaoning Province, there are also large magnesite reserves in Shandong, Tibet, Xinjiang, and Gansu (Table 1.6). Table 1.7 shows proven reserves of magnesite minerals in China from 2015–2017.

Table 1.7 Proven reserves of magnesite minerals in China from 2015–2017

Mineral	Unit	2015	2016	2017	Forecast resources
Magnesite	Ore, 100 million tonnes	29.7	30.9	31.15	131.4

Table 1.8 Global magnesite output (1000 tonnes) in the period of 2013–2014

Country	2013	2014
U.S.A	10	5
Australia	130	130
Austria	220	200
Brazil	140	150
China	4900	4900
Greece	100	115
India	60	60
Korea	70	80
Russia	370	400
Slovakia	200	200
Spain	280	280
Turkey	300	300
Other countries	130	150
Global total	6910	6970

For various reasons related to policies, the output of magnesite in other countries is far less than that in China. Table 1.8 shows the global magnesite outputs (1000 tonnes) in the period of 2013–2014.

Magnesite is the main raw material used for smelting metal magnesium. It is a carbonate mineral with a trigonal system. Its molecular formula is $MgCO_3$, and its theoretical composition in mass fractions is: 47.81% MgO and 52.19% CO_2. Magnesite minerals have two structures: an amorphous structure and a crystalline structure. The former has no luster, and the latter belongs to the hexagonal system, which has a glassy luster. The color of magnesite is mostly white or light yellow and sometimes light red, but the color is brown when the magnesite contains iron. When magnesite is used as a raw material for magnesium smelting, either an electrolysis method or silicothermic method can be used. Magnesite can also be used as a refractory material, building material, and chemical raw material.

Magnesite belongs to the calcite group and is a carbonate mineral. Its main component is $MgCO_3$, and it often contains impurities such as $CaCO_3$, $FeCO_3$, $MnCO_3$, Al_2O_3, and SiO_2. With the presence of impurities, magnesite often turns into calcium magnesite, iron magnesite, manganese magnesite, aluminum magnesite, silica magnesite, and so on. Magnesite crystal is rare. It belongs to the trigonal system. Magnesite can be divided into two types according to its mineral characteristics: crystalline magnesite and amorphous magnesite. The aggregates are usually dense blocks or grains, exhibiting grayish white, white, light red (containing Co), or yellowish brown (containing Fe) colors. Their density is in the range of 2.9–3.1 g × cm^{-3}, and their hardness is in the range of 3.5–4.5.

At present, the proven reserves of magnesite in the world are about 13 billion tonnes. China's total reserve of magnesite accounts for 1/4 of the world's total, with the proven reserves reaching 3.1 billion tonnes and the retained reserves being 3.0 billion tonnes. Both the proven reserves and retained reserves of China rank no. 1 in the world [12]. The reserves of other major magnesite-producing countries are: the former Yugoslavia: 14 million tonnes; Greece: 30 million tonnes; Brazil: 40 million tonnes; North Korea: 3 billion tonnes: Canada: 60 million tonnes; the United States: 70 million tonnes; Austria: 80 million tonnes; India: 100 million tonnes; the Czech Republic: 500 million tonnes; New Zealand: 600 million tonnes; the former Soviet Union: 2.2 billion tonnes [13]. In China, 27 magnesite deposits distributed in 9 provinces and autonomous regions have been identified. A large portion of reserves is concentrated in Laizhou, Shandong Province (286 million tonnes) and the southern part of Liaoning Province (2.569 billion tonnes). The aggregate reserves of these two regions are 2.855 billion tonnes, accounting for 95.2% of the country's total. In contrast, the aggregate reserves of Sichuan, Qinghai, Tibet, Anhui, Gansu, Xinjiang, and Hebei are only 145 million tonnes, accounting for 4.8% of the country's total. The major magnesite production firms are located in Yexian County (Shandong Province) and Dashiqiao and Haicheng (Liaoning Province). As China is the biggest producer and exporter of magnesite resources in the world, there is a strong demand for China's magnesite in the international market. China's magnesite-producing firms have a strong competitive advantage in the international even through their magnesite production processes are not very advanced. However, we should also have a clear understanding that China still lags far behind some developed countries in the utilization of magnesite resources, wasting a lot of ore resources [14].

Magnesite deposits can be divided into four types: hydrothermal metasomatism, metamorphosed sedimentary, vein filling, and weathered residual. The most important industrial type of deposit is the metamorphosed sedimentary type, which is also the type that is most intensively exploited by domestic and foreign firms. Deposits of this type are bigger than the other types, with reserves ranging from a few millions tonnes to hundreds of millions tonnes. Moreover, the deposits are mostly planar or have the shape of a lens, with dozens of layers. The ore quality is excellent, and the content of MgO is in the range of 35–47%. The magnesite resources in China are characterized by shallow burial, good quality, and large-scale deposits, and the proportion of carbonates can reach 96%. Among the 27 magnesite mining areas in China, there are 11 large-scale deposits with a reserve equal to or larger than 50 million tonnes. The aggregate reserve of these mining areas accounts for 95% of the country's total. Relevant data show [15] that among the retained reserves of magnesite mineral, ores of high quality (super class and class 1) account for more 37.58%. Table 1.9 shows the compositions of ores mined from major magnesite mineral areas in China.

It can be seen from the data in the above table that once the magnesite ores in the magnesite mines in Liaoning, Shangdong, Sichuan, and other provinces are mined,

Table 1.9 Compositions of ores mined from major magnesite mineral areas in China (%)

Component		SiO_2	Al_2O_3	Fe_2O_3	CaO	MgO	Loss on burning
Liaoning	Haicheng super class	0.17	0.12	0.37	0.50	47.30	51.13
	Haicheng Xiafangshen	0.26	0.06	0.27	0.45	47.30	50.99
	Dashiqiao class-1 mine	1.90	0.47	0.50	1.14	45.80	48.87
	Huaidong section of Qingshann Mine	0.66	–	–	0.73	46.91	–
	Yingkou Class-1 mine	1.13	0.21	0.33	0.33	47.14	50.97
Shandong	Yexian West Mine class-1 Mine	0.90	0.18	0.55	0.37	47.00	51.11
	Yexian West mixed class	4.95	1.39	0.93	0.86	44.08	47.33
	Yexian East mixed class	3.87	0.59	0.58	0.75	46.43	48.21
Sichuan	Ganluo Yandaii	0.24	–	–	4.30	44.41	–
	Hanyuan Guidai	0.10	–	–	0.80	46.91	–
Dahe, Xingtai, Hebei Province		0.30	–	–	3.94	42.53	–
Biegai, Subei, Gansu		0.25	–	–	4.58	43.81	–

they can be used as MgO after roasting. In China, most of the magnesite mines are operated in the open-pit mining mode. For the magnesite ores mined from the Haicheng and Dashiqiao magnesium mines, the electrolysis method can be used for magnesium smelting. Haicheng magnesium mine is located in Pailouling, which is southeast of Haicheng City, Liaoning Province. In the mining area, there are super high-quality magnesite deposits in the Xiafangshen, Jinjiapu, and Wangjiapu areas. This mining will last 25 years, and the ores of class-1 and super class quality account for 50–55% of the total. The production capacity of this magnesite mine is 1.7 million ton/year, the stripping amount is 2.8 million ton/year, the total annual excavating and stripping amount is 4.5 million ton/year, the average stripping-to-excavating ratio is 0.85, the mining recovery rate is 92%, the annual labor productivity of mining workers is 5948.9 ton/person, and the unit mining cost is 26.73 yuan/ton.

3. **Carnallite**

Carnallite is an aqueous complex salt composed of $MgCl_2$ and KCl, and its molecular formula is $KCl \cdot MgCl_2 \cdot 6H_2O$. Theoretically, the contents of $MgCl_2$, KCl, and H_2O are 34.5%, 26.7%, and 38.8%, respectively. The molar ratio of $MgCl_2$-to-KCl is 1. Carnallite belongs to the orthorhombic system. Pure carnallite is white. Because natural carnallite minerals contain impurities, such as NaCl, NaBr, $MgSO_4$, and $FeSO_4$, they exhibit various colors, including pink, yellow, gray, and brown. The hardness of carnallite is 1–2, and the specific gravity is 1.62 g $\times$ cm^{-3}. The largest carnallite deposits in the world are located in the Urals of the former Soviet Union and the Elbe area of eastern Germany. There is also a large amount of high quality carnallite in Qinghai Salt Lakes of China [16].

4. Serpentine

Serpentine, whose chemical formula is $Mg_3Si_2O_5(OH)_4$, is composed of silicon oxygen tetrahedra and magnesium hydroxide octahedra layer-by-layer at a ratio of 1:1. One layer of magnesium hydroxide octahedra and one layer of silicon oxygen tetrahedra form a crystal layer [17]. Theoretically, the contents of H_2O, SiO_2, and MgO are 12.9%, 44.1%, and 43%, respectively. Because the serpentine ore usually contains a small amount of oxides of Al, Fe, Ni, and Ca, its chemical composition and the theoretical contents of chemicals in it vary from one deposit to another and even vary from one section to another section of the same deposit.

China has rich serpentine resources. With more than 1.5 billion tonnes of proven reserves distributed across the country, China has an obvious advantage in exploiting serpentine [18]. Among all serpentine resources, the reserves in the western region account for 98% of the country's total (only the reserves in the western and eastern Mangya mining areas in Qinghai Province claim a share of 48%), and the reserves in the remaining regions only account for 2% of the country's total. By provinces, Qinghai Province boasts the largest reserves, accounting for 63% of the country's total, followed by Sichuan Province (20%) and Shaanxi Province (12%). These three provinces combined account for 95% of China's serpentine reserves.

5. Liquid mineral resources (magnesium salt resources in salt lakes of China)

China's salt lakes containing magnesium salts are mainly distributed in the northern part of the Tibet Autonomous Region and Qaidam Basin of Qinghai Province. The reserves of magnesium salts in Qaidam Basin account for 99% of the total proven magnesium salt reserves in China. Magnesium salt resources in the Qaidam Basin are mainly distributed in Chaerhan Lake, Yiliping Lake, East Taijinar Lake, West Taijinar Lake, Dalangtan Lake, Kunteyi Lake, Mahai Lake, and other salt lakes. The salt lakes of Qarhan, Yiliping, East Taijinaer, and West Taijinaer have only magnesium chloride, whereas Dalangtan, Kunteyi, Mahai, Dachaidan, and other mining areas have basically equal amounts of magnesium chloride and magnesium sulfate. The total proven reserves of magnesium chloride are 4.281 billion tonnes (including 1.908 billion tonnes of basic reserves), and the retained reserves are 4.070 billion tonnes (including 1.798 billion tonnes of basic reserves). The total proven reserves of magnesium sulfate are 1.722 billion tonnes, including 1.229 billion tonnes of basic reserves.

Liquid mineral resources mainly exist in underground brine, salt lake water, and sea water. Containing 2×10^{15} tonnes of magnesium, sea water is the largest repository of magnesium mineral resources. Each year, large amounts of $MgCl_2$ and $MgSO_4$ are extracted from underground brine, salt lake brine, and seawater. Salt lakes usually refer to lakes that have a salt content higher than 35%. A salt lake is a place where various mineral resources concentrate. Most of China's salt lakes are distributed in

Qaidam Basin, Qinghai Province, the northern Tibet Autonomous Region, Shanxi Province, and Gansu Province. The salt lakes in these regions account for 99% of the identified magnesium salt resources in China. Among these salt lakes, Yuncheng Salt Lake in Shanxi Province contains 6.5×10^9 tonnes of magnesium salts, Qarhan salt lake of Qinghai Province contains more than 3.0×10^9 tonnes, and the salt lake deposits in Gansu Gaotai County contain 2.9×10^6 tonnes [19]. Magnesium salt resources in the Qaidam Basin are mainly distributed in Chaerhan Lake, Yiliping Lake, East Taijinar Lake, West Taijinar Lake, Dalangtan Lake, Kunteyi Lake, Mahai Lake, and other salt lakes. Among magnesium salts, magnesium chloride and magnesium sulfate are the two-dominant species. The salt lakes of Qarhan, Yiliping, East Taijinaer, and West Taijinaer have only magnesium chloride, whereas Dalangtan, Kunteyi, Mahai, Dachaidan, and other mining areas have basically equal amounts of magnesium chloride and magnesium sulfate. The total proven reserves of magnesium chloride are 4.281 billion tonnes (including 1.908 billion tonnes of basic reserves), and the retained reserves are 4.070 billion tonnes (including 1.798 billion tonnes of basic reserves). The total proven reserves of magnesium sulfate are 1.722 billion tonnes, including 1.229 billion tonnes of basic reserves [20].

The brine in Yuncheng Salt Lake belongs to the Na^+, M^{2+}/Cl^-, and SO_4^{2-}-H_2O quaternary system. The lake is a sulfate-type compound salt lake, containing 9.3 million tonnes of magnesium salts. For a long time, the main product of the mining activity at Yuncheng Salt Lake has been anhydrous sodium sulfate. At present, the annual output has reached 3.6 million tonnes, accounting for more than 30% of the country's total.

At present, all of the large amount of high-quality magnesia used in the production facilities in China is extracted from brine. For example, there are 31 salt lakes containing potassium and magnesium salts in Qinghai Province. Among them, the Chaerhan Salt Lake located in the middle and eastern Qaidam Basin is the largest and shows a trend of drying up. The lake hosts a modern sedimentary deposit mainly composed of liquid potassium and magnesium salts. The mining area is 5856 km², and the altitude is 2677–2680 m. The reserve of intercrystalline brine is as large as 6.73 billion m³, belonging to the $NaCl$-KCl-$MgCl_2$-H_2O quaternary system. It is now in the deposition period of potassium carnallite ($KCl \cdot MCl_2 \cdot 6H_2O$) and bischofite ($MgCl_2 \cdot 6H_2O$). Carnallite and bischofite can be obtained from brine through natural freezing and exposure to sunshine. They can be used as raw materials for magnesium smelting. At the end of 1992, an industrial test of the smelting magnesium process with potassium carnallite serving as the raw material started in Qinghai Minhe Magnesium Plant was conducted. After the test was successful, the process was applied in industrial production, using carnallite as the raw material. The carnallite is composed of KCl (17–20%), $MgCl_2$ (28–30%), $NaCl$ (10–15%), and $CaSO_4$ (85%); 25 tonnes of carnallite is needed to produce 1 ton of metal magnesium.

Using magnesium salts from salt lakes to produce metal magnesium is also an important development direction, as exemplified by the metal magnesium smelting plant (under construction) of Qinghai Salt Lake Industrial Group whose production capacity will be 400,000 ton/year. The magnesium smelting processes include the Pidgeon process (silicothermic reduction method) and the electrolysis method. At present, most magnesium smelting plants use the Pidgeon process. In 2010, the output of metal magnesium produced by the Pidgeon process in Shanxi Province reached 380,000 tonnes, accounting for 60% of the country's total.

At large salt lakes in the United States, bischofite is extracted to produce a series of magnesium salts, including powdery magnesium oxide, high-purity magnesium hydroxide, light burned magnesium oxide, and heavy burned magnesium oxide, forming an industrial chain of magnesium products. In the 1980s, Yamaguchi Company of Japan developed a technique for preparing magnesium sulfate whiskers. China has also done a lot of work in exploiting magnesium halide resources in salt lakes and developed processes to produce magnesium chloride hexahydrate, potassium magnesium sulfate fertilizer, ordinary magnesium sulfate, ordinary magnesium hydroxide, and flame retardant magnesium hydroxide. However, most of them are low value-added traditional products. There are few reports on the industrialization of high value-added products such as high-purity magnesium sulfate and flame-retardant magnesium hydroxide.

China is a top possessor of magnesium mineral resources in terms of both quantity and quality. This gives China a unique advantage in the exploitation and utilization of magnesium mineral resources [21], and thus, there are very favorable conditions for developing magnesium and magnesium alloy products in China. At the same time, there are still some problems in the exploitation and utilization of magnesium mineral resources in China, and these are mainly manifested in the following aspects [22].

First, although the total reserve of magnesium resources of China is enormous, the per capita resource is small, and the resources are unevenly distributed in different regions [23]. Among the magnesium resources, the proven reserves are less than the controllable reserves, and the consumption rate of magnesium resources will gradually exceed the growth rate of reserves, as indicated by the declining retained resources in most mining areas. However, the number of newly discovered magnesium mineral deposits is decreasing gradually, resulting in a shortage of backup reserves.

Second, most of the firms producing metal magnesium and magnesium alloys in China are small in scale; their production equipment is not sophisticated, and their technologies are not very advanced. Consequently, their production activities often cause serious water and air pollution. At present, firms producing metal magnesium and magnesium alloys in China mainly produce low value-added products, and low technology content is therefore still trapped in the development mode of "survival at the cost of environment and resources". China has yet to fully exploit its advantage of magnesium mineral resources.

Third, there is a serious problem of overmining magnesium resources, resulting in a low utilization level of magnesium resources. The problem is most pronounced in the production and utilization of magnesite and dolomite: there is a lack of classification systems for more efficient utilization of the ore resources of the same kind.

Fourth, mining activities inflict serious damage to the ecological environment in mining areas. When the stratum containing the dolomite is older, the content of MgO in the ore is higher. Dolomite rocks that are hard can therefore withstand weathering better than many other rocks. So, they are usually found at high positions, and their mining process usually begins with large-scale blasting. Differing from dolomite, most magnesite and serpentine deposits are shallow, which is suitable for large-scale open-pit mining. Generally speaking, because of the lack of supervision and management of mining activities, mining of magnesium ores in a mining area is usually accompanied by damage to a large area of surface vegetation, inflicting serious damage to the ecological environment and affecting the ecological balance.

Fifth, there is a serious workplace safety problem in mining areas. In the process of ore mining, the operation is often carried out in a disordered manner as a result of the lack of planning, giving rise to safety accidents. Therefore, it is necessary to take measures to standardize the operation, strengthen supervision, and improve planning to achieve the best results in the exploitation and utilization of magnesium resources.

1.3   Development History of the Magnesium Industry

1.3.1   Development Course of the Magnesium Industry

From 1808 to the present, the development of the magnesium industry has continued for more than 200 years, and the industrial production era of magnesium has spanned more than 130 years since 1886. Before the 1950s, the development of the magnesium industry mainly relied on the military industry. Demand by the military industry, especially with the outbreak of two world wars, was a significant stimulus for the world's magnesium output. Magnesium output increased significantly during war and fell at the end of war. In 1910, the world's magnesium output was about 10 tonnes. In 1914, World War I broke out, and the world's output of primary magnesium increased to 350 tonnes the following year. By the end of the war in 1917, the world's output of primary magnesium had boomed to 3000 tonnes. With the end of the war, the world's annual output of primary magnesium fell to 330 tonnes in 1920. Similarly, under the stimulus caused by World War II, the world's magnesium output increased to 32,000 tonnes in 1939 and reached a peak of 235,000 tonnes in 1943. At the end of the 1940s, the world's annual magnesium output dropped again.

Roughly speaking, the development of the magnesium industry has gone through three stages. The first stage is the chemical method stage. British chemist Humphrey Davy electrolyzed a mixture of magnesium oxide and mercury and obtained magnesium amalgam as an electrolytic product. After removing mercury from the magnesium amalgam via distillation, silver-white metallic magnesium was obtained; this was the first appearance of magnesium in its elemental form, but the output was very small [24]. By 1831, French scientist Antoine Alexandre Brutus Bussy used molten anhydrous magnesium chloride ($MgCl_2$) and potassium vapor as raw materials to initiate a reduction reaction that produced a large amount of metallic magnesium in the laboratory for the first time [24]. Scientists in Britain, the United States, and other countries did not begin to use chemical methods to prepare metallic magnesium until the 1860s, and a slight increase in the output of metallic magnesium was achieved. This stage is considered to be the first stage of the development of the magnesium industry, specifically, the chemical method stage. Magnesium production in this stage was generally limited to small-scale laboratory production and industrial production did not emerge. This stage lasted for 78 years until the world's first magnesium plant was completed in 1886.

In 1833, British scientist Michael Faraday achieved the first electrolytic reduction of molten $MgCl_2$ to produce metallic magnesium. In 1852, German chemist Robert Bunsen studied the use of electrolysis for producing metals. He was the first to successfully produce metallic magnesium via electrolysis of molten $MgCl_2$, and he established the world's first electrolytic cell and used it for the electrolysis of anhydrous $MgCl_2$. In 1886, Germany established the first industrial-scale electrolysis cell and implemented industrial production of magnesium. Griesheim-Elektron constructed the world's first facility for commerciall production of magnesium in Stassfurt. In 1860, Johnson Matthey and CoinMna-Chester began to use similar processes to produce magnesium. In 1896, British Chemische-Fabrik Griesheim-Elektron and Aluminum Magnesium Fabrik jointly purchased the electrolytic magnesium process. Until 1914–1915, these companies remained the world's most important manufacturer of magnesium. In 1916, the US DOW Chemical Company established a magnesium smelting plant that was the world's largest magnesium company at that point, making DOW the world's leading manufacturer of magnesium. This time period is considered to be the second stage of the development of magnesium industry and is called the electrolytic smelting stage. As the demand for magnesium in market increased on a year by year basis along with continuous improvements to the electrolysis process and equipment, the electrolytic smelting method to produce magnesium has been one of the world's leading processes [3]. More than 80% of magnesium in developed countries is produced via electrolytic smelting.

The structure of electrolytic cell for magnesium has changed a lot since the magnesium electrolytic cell was used in industrial production at the end of the nineteenth century. The original electrolytic cell was a simple electrolytic cell without a diaphragm. Since the 1930s, this electrolytic cell has included a diaphragm. Since the 1960s, a new type of magnesium electrolytic cell without a diaphragm emerged and has been used, and this cell promoted the magnesium industry into a new stage of development. Until the mid-1990s, production of magnesium via electrolysis has always been dominant approach, and its output accounted for 70–75% of the total magnesium output.

As people's demand for magnesium and magnesium alloys continues to increase, the production of metallic magnesium via electrolysis alone cannot sufficiently meet people's needs any more. Several scientists have used chemical methods to study the smelting of magnesium-containing ore via thermal reduction. Therefore, the third stage in the development of the magnesium industry is called the thermal reduction stage. Because the application of magnesium and its alloys have gradually been extended, the demand for magnesium has increased accordingly. In 1913, the vacuum thermal reduction method was used for the process of reducing magnesium oxide to make magnesium. The first method using silicon as a reducing agent to reduce magnesium oxide to produce metallic magnesium emerged in 1924. The use of silicon aluminum alloy as a reducing agent to reduce magnesium oxide to produce magnesium was achieved in 1932. In 1941, Professor L. M. Pidgeon of the University of Toronto in Canada successfully established a pilot plant in Ottawa where ferrosilicon was used as a reducing agent to extract magnesium in a vacuum reduction tank [25]. The silicothermic smelting process for magnesium is named after Professor L. M. Pidgeon and called the Pidgeon process. The Pidgeon process and the electrolysis process are still the two main magnesium smelting methods. In the early 1950s, a new thermal reduction smelting technology for magnesium was proposed by the French Pedmey Electric Metallurgical Company. This process uses ferrosilicon or aluminum as a reducing agent to reduce dolomite, and then magnesium is obtained in a large internally-heated furnace. This process is called the magnetherm method and is a semicontinuous method. The smelting method underwent rapid development in the middle of the twentieth century, and the amount of primary magnesium produced via this method accounts for one-half of the total output of primary magnesium in industrially developed countries. Since the early 1970s, the magnetherm method has been used and promoted in France, the United States, the former Yugoslavia, and Brazil. In China, magnesium is mainly produced through the Pidgeon process, and it has gradually become a advanced technological methods used in the production of magnesium industry today. In the late 1980s, with the further development of the Pidgeon method in China, the electrolytic smelting process was gradually replaced in the field of magnesium production. For instance, several magnesium plants that were planned by the U.S. Dow Chemical Company, Bosglon in Norway, Noranda Company in Canada, Puji Company in France, and Northwest Alloy Plant of the United States and Australia have stopped construction or have been postponed. Among the production enterprises currently operating abroad, the large-scale plants (with a capacity around 50,000 t/year) still use the electrolytic method. For example, the American Magnesium Company has a production capacity of 43,000 tonnes per year, and the Israeli Dead Sea magnesium plant has a production capacity of 55,000 tonnes per year. In addition to Commonwealth of Independent States (CIS) countries, the current foreign metal magnesium production capacity is 100,000–120,000 tonnes per year, and the products they produce still have a wide international market. The advantages and disadvantages of the two magnesium smelting methods are compared (Table 1.10).

1.3.2   Worldwide Distribution of Magnesium Industry (Outside of China)

Since 1960, various excellent properties of magnesium and its alloys have been gradually discovered, and the use of magnesium in the civilian market has greatly promoted the development of the magnesium industry. Magnesium has been widely used as an

Table 1.10 Comparison of two magnesium smelting method

Items	Electrolysis	Silicothermic (Pidgeon process)
Principle	The solution containing $MgCl_2$ is dehydrated to form anhydrous $MgCl_2$. The electrolysis of anhydrous or molten $MgCl_2$ produces metallic magnesium	Carbonate ore is calcined to produce magnesium oxide, and ferrosilicon is used for thermal reduction to produce metallic magnesium
Raw materials	Brine, magnesite, carnallite, etc	Dolomite
Advantages	Energy savings, good product uniformity, continuous production process	Low investment in equipment, low technical difficulty, high purity magnesium and abundant resources
Disadvantages	The preparation of anhydrous $MgCl_2$ is relatively difficult to control; dehydration requires higher temperature and an acidic atmosphere, has a higher energy consumption, causes prominent corrosion on equipment, and the treatment cost for "waste water, waste gas and waste residue." are large	Low heat utilization rate, short life of reduction furnace, discontinuous production process
Portion of production capacity	About 20%	About 80%

additive in materials such as aluminum-based alloys, architectural aluminum profiles, and beverage cans. Meanwhile, the continuous development and progress of magnesium production technology has expanded the production scale, increased output, and reduced energy consumption and costs.

After 1970, metallic magnesium began to be used for desulfurization in the steel-making process and became one of the main desulfurizers in the pre-treatment of hot metal. At the same time, for the reason of light weight, pressure castings made by magnesium and magnesium alloy have begun to be used in cars to reduce the weight and energy consumption of vehicles [26]. Since 1990, developments in magnesium-containing composite materials and ultralight magnesium-lithium alloys have further expanded the application scope of magnesium, spreading to include almost all industrial fields.

After entering the twenty-first century, the application and promotion of magnesium in various fields have gradually stabilized, and thus the world's primary magnesium output tended to increase steadily. China and Russia possess the largest magnesium processing equipment. These two countries produce two-thirds of the world's magnesium oxide. Japan, the Netherlands, and the United States mainly extract magnesium from seawater and brine. Their output of magnesium oxide accounts for about 52% of the total global magnesium output from seawater and brine. Australia,

Brazil, China, Iran, Israel, Japan, South Korea, Mexico, Norway, Russia, Turkey, the United Kingdom, and the United States also produce electrically-fused magnesia. In 2012, the global production capacity of fused magnesia increased by 175,000 tonnes, and the annual global production capacity of dead-burned magnesia was approximately 8.5 million tonnes.

According to data released by the US Geological Survey in 2015, the global output of primary magnesium in 2014 was 907,000 tonnes, year-on-year rises is 29,000 tonnes. China's output of primary magnesium is the highest in the world, accounting for 88.2% of the total world output. Besides China, the major producers of magnesium are Israel (30,000 tonnes), Russia (28,000 tonnes), and Kazakhstan (21,000 tonnes).

The main foreign magnesium manufacturers and their outputs are shown in Table 1.11. Table 1.12 shows the world's main magnesium-producing countries and their outputs.

Although the world's magnesium industry has developed rapidly, the production and development of primary magnesium has been unbalanced. After the 1990s, because of the impacts of the rapid increases in the output and export volumes of China's magnesium industry [27], some countries (such as France) have had to close a number of magnesium smelters, and in some countries such as Canada and Australia, it was difficult to construct new magnesium plants or to put them into normal production. The magnesium plant of Japan's Udu Kosan Co., Ltd. also withdrew from the world's magnesium smelting industry [28]. At present, apart from China, the main

Table 1.11 Major foreign magnesium manufacturers and productions

Nations	Company name	Annual production capacity (10,000 tonnes)
Israel	Dead Sea Magnesium Co., Ltd	3.3
Russia	Solsmck magnesium	2
Russia	Avisma magnesium	1.5
Brazil	Rima magnesium	2
Kazakhstan	Magnesium plant in Kamennogorsk	0.5
Norway	Magontec Xi'an Co., Ltd	0.5
Canada	Canada Magnesium Corporation	1.25
Egypt (Australia Magnesium)	Magnesium smelting plant in Sokhna	20
Congo	Kouilou magnesium smelting plant	6
USA	Rowley magnesium plant	5.9–7.3
Australia	Australia Magnesium International Ltd	7.1–20

Table 1.12 World's leading magnesium-producing nations and their outputs in 2013 and 2014 (10,000 tonnes)

Nations	2013	2014
Brazil	1.6	1.6
China	77	80
Israel	2.8	3.0
Kazakhstan	2.3	2.1
Korean	0.8	1.0
Malaysia	0.1	0
Russia	3.2	2.8
Global output	87.8	90.7

countries that producing magnesium metal are Israel, Russia, Kazakhstan, Brazil, etc.

1.3.3 Development of Magnesium Industry in China

Before 1995, China's primary magnesium output was very small, and the primary magnesium outputs of western countries accounted for about 80% of the world's total magnesium output. In the 1980s, there were only three magnesium smelters in China, Baotou Guanghua Magnesium Group Company, Qinghai Minhe Magnesium Plant, and the magnesium branch of Lushun Aluminum Plant. All three magnesium smelters use electrolytic methods, and thus, the investment cost of building a plant is relatively high and the preparation of anhydrous magnesium chloride is difficult to control. Also, the production is difficult, the equipment is easily corroded, and the output is low. Since 1987, the Pidgeon process has undergone rapid development in China, and many small magnesium smelters that use the Pidgeon process were built during this period [29]. The specific reasons for this include the following: the Pidgeon method is a relatively simple smelting process, and the equipment selection is simple; the investment in the construction of the plant is small; the purity of the magnesium produced is high; the production scale is flexible, and dolomite-rich mineral resources in China can be used directly as raw materials. In particular, the Pidgeon process was put forward in 1941, and after years of exploration and improvement, it is more in line with China's actual national conditions.

Since 1999, China has gradually become the world's largest producer of magnesium products. By 2007, China's output of primary magnesium reached 624,700 tonnes, of which the output of magnesium alloy reached 226,200 tonnes, as shown in Table 1.13. The magnesium industry has developed rapidly in China because of the advantages of resources, energy, labor and production methods. In 2007, China's output of primary magnesium accounted for more than 80% of the world's total output.

Table 1.13 Output of Mg and magnesium alloys in China from 2001 to 2007 (numbers are in 10,000 tonnes)

Year	2001	2002	2003	2004	2005	2006	2007
Global magnesium output	47.86	52.41	49.08	63.34	65.78	70.87	77.67
China's magnesium output	19.97	32.50	34.18	44.24	45.08	51.97	62.47
Global annual growth rate (%)	–	9.51	−6.35	29.05	3.85	7.74	9.60
China's annual growth rate %	–	62.74	5.17	29.43	1.90	15.28	20.20
Portion of China's output (%)	41.73	62.01	69.64	69.85	68.53	73.33	80.43
China's output of magnesium alloys	–	–	–	13.58	17.51	21.10	22.62
Portion of magnesium alloys in magnesium output (%)	–	–	–	30.70	38.84	40.60	36.21

Table 1.14 Mg output in China from 2004 to 2009 (numbers are in 10,000 tonnes)

	Production capacity of primary magnesium	Output of primary magnesium	Output of magnesium alloy	Output of magnesium powder
2004	76.0	45.0	13.5	9.1
2005	81.5	46.8	17.5	8.6
2006	90.2	52.6	21.2	10.0
2007	97.7	65.9	22.6	11.1
2008	116.2	63.1	21.1	13.9
2009	131.9	50.2	16.4	11.1

Based on statistics provided by the National Bureau of Statistics and the Magnesium Branch of China Nonferrous Metals Association, the year-on-year changes in production capacity and output of metallic magnesium in China from 2004 to 2009 are listed in Table 1.14. In 2009, 64 magnesium smelters were counted. The production capacity of primary magnesium was 1.319 million tonnes shows a year-on-year increase of 13.52%. The increase in 2009 was 5.34% lower than that in 2008. The output of primary magnesium was 502,000 tonnes and shows a year-on-year decrease of 20.44%. The output of magnesium alloys was 164,000 tonnes, showing a year-on-year decrease of 22.5%, and the output of magnesium powder was 111,000 tonnes, showing a year-on-year decrease of 19.8%. The decreases are a result of the financial depression that occurred from the end of 2008–2009; the magnesium industry market shrank, and the coal, steel, and alloy industries experienced downturns. These negative factors caused most of the magnesium producers to cut or suspend production, leading to a rapid decline in magnesium production and to greater gaps in magnesium production capacity and output.

After 2010, because of improvements in the world economic situation, the magnesium output in China showed a recovery. In the first half of the year, the output of primary magnesium was 324,500 tonnes, which was an 86.81% increase compared to the same period in previous year. In 2010, data for newly-built magnesium smelting enterprises in China are shown in Table 1.15. Of the manufacturers listed in Table 1.15, all of them (except Qinghai Salt Lake Group Magnesium Industry Co., Ltd., which uses the Pidgeon process) use electrolysis to smelt magnesium.

According to statistics from the China Nonferrous Metals Industry Association, China's output of primary magnesium in 2014 was 873,900 tonnes, which is a 13.53% increase compared to the same period the previous year. China's primary magnesium production area has also spread from the original Hubei and Henan provinces to coal-rich provinces, such as Shaanxi, Shanxi, and Ningxia. Shaanxi is the province with largest magnesium output and produced a total of 404,600 tonnes in 2014, accounting for 46.30% of the national output. Of the areas in Shaanxi, the cumulative production of the Yulin area was 396,300 tonnes. Fugu county in the Yulin area has an output of 348,100 tonnes, which accounts for 39.83% of the country's total magnesium output and about 86% of the province's magnesium output. The year-on-year changes of my country's primary magnesium-producing areas in 2014 are shown in Table 1.16.

In 2014, the primary magnesium output of Shaanxi Province ranked first in China. Shaanxi Province's output reached 404,600 tonnes, which accounts for 46.30% of the total national output, and this province remained the largest primary magnesium-producing area in China for two consecutive years.

According to the latest statistics provided by the General Administration of Customs, China's total exported magnesium was 435,000 tonnes in 2014, showing a year-on-year increase of 5.80%. The export of magnesium ingots was 227,300 tonnes, showing a year-on-year increase of 7.18%. The export of magnesium alloy was 106,500 tonnes, showing a year-on-year increase of 4.42%. The export of magnesium powder was 88,000 tonnes, showing a year-on-year increase of 3.05%. The export of magnesium waste and scraps were 2900 tonnes, showing a year-on-year increase of 87.66%. The export of processed magnesium materials was 3700 tonnes, showing a year-on-year decrease of 16.83%. The export of magnesium products was 6600 tonnes, a year-on-year increase of 15.65%. In 2016, the primary magnesium output in China reached 910,300 tonnes; this was the maximum magnesium output in the past ten years [30]. In 2017, the global output of magnesium metal exceeded 1.2 million tonnes. China is one of the largest producers of magnesium in the world, with China's magnesium output accounting for more than 85% of the world's total output; the magnesium production and export volume rank first in the world. All exported magnesium is produced via the Pidgeon method [31]. In addition to being the leader of primary magnesium production, China has developed deep magnesium processing technology Besides the leading position in the production of primary magnesium, deep processing technology of magnesium has been gradually transferred to China. Thus, it is anticipated that the magnesium industry in China has a bright future.

From January to June 2017, the export of China's magnesium and its products (including waste and scraps) was 245,083 tonnes, which is a 52.8% increase compared to the same period in the previous year. From January to June 2017, the value of China's exported magnesium and magnesium products (including waste and scraps) was 566,948 thousand USD, which is a 47.6% increase compared to the same period in the previous year. The export statistics of China's magnesium and

Table 1.15 Newly-built Mg manufacturers in 2010 (numbers are in 10,000 tonnes)

Area	Name	Location	Scale under construction	Scale planned	Owner	Time to start production
Ningxia	Ningxia Sun Magnesium Industry Co., Ltd	Sun Mountain Development Zone, Wuzhong, Ningxia	3.5	10.0	DunAn Holding Group Co., Ltd	2010
	Ningxia Huaying Mining Co., Ltd	Sun Mountain Development Zone, Wuzhong	1.0	5.0	Ningxia Huaying Mining	2010
	Ningxia Kaitai Magnesium Industry Co., Ltd	Huianbao town, Yanchi county	1.5	5.0	Shanghai Zhonghe	2011
Inner Mongolia	Baotou Dongfang Ecological Magnesium Industry Co., Ltd	Guyang county, Baotou	1.0	5.0	China Direct Investment Corporation	2011
Shanxi	Shengying He Light Metal (Shanxi) Company	Panlong Town Economic Park, Wuxiang	1.0	5.0	China Minmetals Corporation	2010
Shaanxi	Fugo County magnesium industry group	Fugo County	1.5	5.0	Fugu Magnesium Group	2010
	Fugo County magnesium industry group	Fugo County	1.5	5.0	Fugu Coal Chemical Group	2010
Qinghai	Qinghai Salt Lake Group Magnesium Industry Co., Ltd	Chaerhan Salt Lake	5.0	40.0	Qinghai Salt Lake Industry Group	2011
Total			16.0	80.0		

Table 1.16 Output variations of main primary magnesium-producing areas in 2014

Areas	Output in 2013 (10 k tonnes)	Output in 2014 (10 k tonnes)	Cumulative year-on-year change (%)
Shaanxi	34.33	40.46	17.86
Shanxi	23.67	24.97	5.49
Ningxia	10.81	9.3	−13.97
Xinjiang	2.29	4.45	94.32
Henan	4.01	4.14	3.24
National	76.97	87.39	13.53

Table 1.17 Statistics for China's volume of exported magnesium and magnesium products (including waste and scraps) from January to June 2017

Month	Weight (ton)	Value (k USD)	Year-on-year ratio in weight	Year-on-year ratio in value
Jan	40,937	100.136	45	49.5
Feb	41,459	94.131	98.7	88.7
Mar	44,407	102.342	46.6	45.6
Apr	41,105	95.610	44.6	38.2
May	36,831	84.195	81.9	70
Jun	40,723	91.563	26	17.1

magnesium products (including waste and scraps) from January to June 2017 are shown in Table 1.17.

From January to July 2019, the volume of China's exported magnesium and magnesium products (including waste and scraps) was 245,083 tonnes, which is a 17.9% increase compared to the same period in the previous year. From January to July 2019, the value of China's exported magnesium and magnesium products (including waste and scraps) was 698,470 thousand USD, which is a 47.6% increase compared to the same period in the previous year. Statistics of China's exported magnesium and magnesium products (including waste and scraps) from January to July 2019 are shown in Table 1.18.

1.4 Production Process of Metallic Mg

At present, more mature methods of producing metallic magnesium in the industrial productions used by various countries are generally divided into two categories according to differences in magnesium ore resources [32–35]. The first category is electrolysis of molten magnesium chloride, in which $MgCl_2$ is used as a molten electrolyte to obtain metallic magnesium via electrolysis with the use of direct current.

Table 1.18 Statistics of China's volume of exported magnesium and magnesium products (including waste and scraps) from 2013 to July 2019

Year	Volume of exported magnesium and magnesium products (including waste and scraps) (ton)	Year-on-year increase in export volume (%)	Value of exported magnesium and magnesium products (including waste and scraps) (k USD)	Year-on-year increase in export value (%)
2013	411,123	10.8	1,187,366	2.6
2014	434,996	5.8	1,171,995	−1.3
2015	405,551	−6.8	1,006,936	−14.1
2016	356,537	−12.1	852,234	−15.3
2017	451,939	26.8	1,052,556	23.5
2018	409,788	−9.8	1,031,498	−2.5
2019 1–7	268,476	17.9	698,470	24.4

Generally, anhydrous $MgCl_2$ or anhydrous carnallite is prepared by chlorinating or dehydrating magnesite, carnallite, brine, or sea water as raw materials, and then magnesium is obtained via electrolysis. The second category is thermal reduction, which is also called the silicothermic method. In recent decades, great breakthroughs have been made in the use of smelting technology for magnesium. Because of the developments in smelting technology, the price of magnesium has dropped significantly, and the magnesium-to-aluminum price ratio has decreased from 1.8 to 1.4 or lower [36].

1.4.1 Magnesium Smelting via Electrolysis

In terms of electrolysis, the technology used for dehydrating magnesium chloride solution to obtain anhydrous magnesium chloride has achieved breakthroughs and has become increasingly mature. Additionally, the structure and capacity of the electrolytic cell have also been undergone great development and improvement, which have reduced the energy consumption of the electrolysis process. Electrolysis of molten magnesium chloride includes two major production processes: production of magnesium chloride and electrolysis of magnesium chloride to obtain magnesium. Electrolytic smelting of magnesium involves electrolyzing anhydrous magnesium chloride in the molten state to generate magnesium and chlorine gas via electrolysis. Depending on the raw materials and processing methods used, electrolysis methods can be mainly divided into the following specific methods [37]: the Dow process [38], Norsk Hydro process, chlorination process of magnesium oxide (IG Farbenindustrie Process) [39], Magnola process [40], carnallite method (Russian Process), and electrolysis of magnesite.

(1) DOW method. In 1916, the American Dow Chemical Company used seawater containing $MgCl_2$ and lime milk as raw materials to extract and prepare $Mg(OH)_2$. $Mg(OH)_2$ is then chemically reacted with hydrochloric acid to form a solution of magnesium chloride. The solution is purified and concentrated to obtain $[MgCl_2 \cdot \frac{3}{2}H_2O]$. $[MgCl_2 \cdot \frac{3}{2}H_2O]$ is used as the raw material for electrolysis and is sent into the electrolytic cell to obtain crude magnesium directly via electrolysis; the chlorine by-product can then be recycled. The temperature for preparing magnesium metal via the Dow Process is generally around 750 °C.

(2) Norsk method. Norway's Norsk company is the main magnesium producer in Europe. It uses brine waste that contains $MgCl_2$ provided by the German potassium industry. Crystal water in $MgCl_2$ crystals is removed via high-pressure dried HCl gas to produce anhydrous $MgCl_2$ powder. Molten $MgCl_2$ is then electrolyzed to prepare metallic magnesium. The Norsk Hydro process is the only method that does not use a chlorination reactor to prepare anhydrous $MgCl_2$.

(3) Chlorination method of MgO (I. G. Farben process). The German IG Farben Industrial Company used natural magnesite and coke as raw materials and calcined them at 700–800 °C to obtain calcined dolomite (MgO) with better activity. MgO power with a size that must be less than 0.144 mm is mixed with carbon, in what is called a briquetting process. The briquettes are calcined and chlorinated in a vertical electric furnace to obtain anhydrous $MgCl_2$, and then $MgCl_2$ is electrolyzed to obtain metallic magnesium. The following chemical reaction occurs during the preparation of $MgCl_2$ (Scheme 1.2)

$$2MgO + 2Cl_2 + C = 2MgCl_2 + CO_2 \tag{1.2}$$

(4) Magnola smelting process. The characteristic of this process is that magnesium chloride is used in a serpentine as the raw material for electrolytic smelting. In this process, the tailings of asbestos minerals were immersed in concentrated hydrochloric acid to prepare $MgCl_2$ solution. The pH and ion exchange technology are adjusted to prepare an ultrahigh pure $MgCl_2$ solution, and then the solution is dehydrated and electrolyzed to obtain crude magnesium.

(5) Carnallite method. The chemical formula of carnallite is $KCl \cdot MgCl_2 \cdot 6H_2O$. In this process, carnallite is purified, crystallized, and dehydrated to obtain anhydrous carnallite, which is then directly electrolyzed to produce metallic magnesium. Because of chlorination during anhydrous treatment, dehydration and hydrolysis reaction of carnallite are weaker than those of $MgCl_2$, showing slight hydrolysis only. At the same time, the electrolytic cell must be cleaned frequently. This is mainly because of the presence of KCl in the carnallite method [41].

(6) Electrolytic magnesium smelting using magnesite as raw material. This method uses magnesite as a raw material to carry out electrolytic magnesium smelting. In this process, magnesite is used as a raw material, and the mineral powders

are chlorinated. The obtained magnesium chloride melt is then electrolyzed to prepare metallic magnesium [42].

In summary, electrolytic smelting processes have advanced production technology and lowered energy consumption, preparing magnesium via electrolysis has two shortcomings. First, the purity of the prepared magnesium is relatively low. Crude magnesium produced via electrolysis reduces the corrosion resistances of magnesium and magnesium alloys because crude magnesium contains chlorides as an electrolyte and some impurities such as Cr, Mn, Fe, Si, Ni, K, and Na. Thus, corresponding measures must be taken to enhance the purity of the prepared metallic magnesium. The second shortcoming is that it is relatively difficult to prepare anhydrous $MgCl_2$. During dehydration of $MgCl_2$ crystals to prepare anhydrous $MgCl_2$, anhydrous $MgCl_2$ is extremely easy to hydrolyze, and it generates basic magnesium chloride, $Mg(OH)Cl$. As a result, the production process of preparing dehydration process is relatively difficult to control. Even under an atmosphere of HCl, the dehydration of hydrated $MgCl_2$ needs to be performed at a high temperature of 450 °C. The difficulty causes a series of problems; specifically, a large investment of production enterprises is required, power consumption is large (production of 1 t of magnesium consumes 12,680–13,250 kWh), resulting environmental pollution is serious, equipment corrodes, and the construction period of plants is long. According to statistics, the cost of dehydrating $MgCl_2$ accounts for more than half of the production cost of magnesium

1.4.2 Magnesium Smelting via Thermal Reduction

Smelting magnesium via thermal reduction uses some reducing agent to reduce magnesium-containing ore under high temperature and vacuum conditions. The formed magnesium vapor is then cooled and crystallized by a condenser to obtain crude magnesium. According to different reducing agents, magnesium smelting via thermal reduction is divided into the silicothermic method, carbon thermal method, and thermal reduction method with carbide. The carbon thermal method and thermal reduction with carbide are rarely used in industry, but the silicothermic method is often used in industry and is divided into an internal heating method and external heating method. The Pidgeon process and Magnétherm process [43] are common methods used to reduce magnesium oxide with ferrosilicon as a reducing agent. The former is an external heating method, and the latter is an internal heating method. According to the continuity of production, the silicon thermal method can be divided into a batch type and semi-continuous type. At present, the Pidgeon process has been widely used to smelt magnesium in China.

In the thermal reduction method, large-scale semi-continuous vacuum reduction furnaces are increasingly being put into production, and computer control has been adopted. Magnesium smelting via thermal reduction is also called as thermal smelting. The principle is that magnesium ore undergoes a reduction reaction in a reduction tank under high temperature and vacuum conditions to form magnesium vapor. The vapor is cooled in a condenser and crystallized to obtain crude magnesium. The reducing agent is generally ferrosilicon. Thermal smelting of magnesium mainly takes place in a reduction tank at 1100–1250 °C and under a vacuum of 1.3–13.3 Pa. 75% Si (Fe) alloy and calcined dolomite are used to carry out reduction reaction to prepare metallic magnesium. The reduction reaction equation is shown as Scheme 1.3:

$$2(MgO \cdot CaO) + Si(Fe) = 2Mg + 2CaO \cdot SiO_2 + (Fe) \tag{1.3}$$

Depending on the different equipment that is used, traditional silicothermic reduction can be divided into four methods: Canada's Pidgeon process, Italy's Balzano process, France's Magnetherm method, and South Africa's Mintek thermal magnesium process (MTMP method).

(1) Pidgeon method. In 1941, L. M. Pidgeon invented the Pidgeon process, using the silicothermic process to smelt magnesium. In this process, ferrosilicon and dolomite (briquettes composed of crushed and calcined dolomite and fluorite) are added in an externally-heated vacuum distillation tank to smelt magnesium. This is an externally-heated intermittent silicothermic-reduction process [44]. Over time, the Pidgeon process has been continuously developed and improved, forming a relatively complete theoretical system. Thus far, the Pidgeon Process has remained the most representative and most widely used magnesium smelting process in China. The Pidgeon process can be divided into five stages: ore crushing, calcination, briquetting, reduction, and refining. During high-temperature calcination of dolomite at 1423–1473 K and 10^{-2}–10^{-1} Torr (1 Torr=133.322 Pa), the following chemical reaction occurs (Scheme 1.4):

$$CaCO_3 \cdot MgCO_3 \xrightarrow{1150-1250\,°C} CaO \cdot MgO + 2CO_2 \tag{1.4}$$

The calcined white powder is crushed, ground, and mixed with fluorite powder (containing 95% CaF_2, mainly acting as a catalyst and without chemical reaction [45] and ferrosilicon powder (75% silicon content) to make briquettes. (The making pressure is 9.8–29.14 MPa). The briquettes are placed in a reduction tank. The reduction reaction takes place in the reduction tank at a temperature of 1190–1250 °C and under a vacuum of 1.3–13.3 Pa to obtain magnesium vapor. The vapor is then condensed and crystallized into crude magnesium in the condenser [46]. The chemical reaction is expressed in Scheme 1.5.

$$2CaO + 2MgO + Si \xrightarrow{1190-1210\,°C} 2Mg + 2CaO + SiO_2 \tag{1.5}$$

After undergoing flux refining, ingot casting, and surface treatment, crude magnesium is finally converted to high-purity metallic magnesium ingots. The flow chart for magnesium smelting using the Pidgeon process is shown in Fig. 1.3.

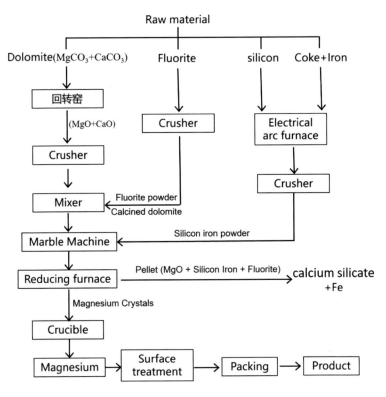

Fig. 1.3 Flow chart of magnesium smelting via the Pidgeon process

The advantages of the Pidgeon process include a low investment, simple production process, short construction period, and high product quality. However, the Pidgeon process cannot proceed continuously. It is an intermittent production method, and each production cycle of the Pidgeon process lasts about 10 h. The entire production cycle can be divided into three stages. The first is the preheating stage. In this stage, dolomite is crushed, placed in a furnace, and preheated to remove CO_2 and moisture in the crushed material. The second stage is a low-vacuum heating stage. In this stage, the reduction tank is sealed and heated under a low vacuum. The third stage is a high-vacuum heating stage. The temperature in the reduction tank is controlled to be about 1200 °C, the degree of vacuum is maintained between 1.3 and 13.3 Pa, and the constant-temperature calcination time is about 9 h. The magnesium vapor generated in the reduction tank condenses and crystallizes on the crystallizer because of the action of the water cooling jacket that surrounds the reduction tank. Finally, the vacuum is terminated, and the reduction tank is opened to remove the crystallized magnesium ring and residue on the condenser.

When applying the Pidgeon smelting process, the preparation of calcined dolomite and briquettes is an important link [47]. First, the Pidgeon process requires calcined dolomite that has higher activity because better quality calcined dolomite is beneficial for the reduction reaction. If the calcination temperature is too high, it will cause the calcined dolomite surface to be over-sintered, and the activity of the calcined dolomite decreases. At the same time, the calcined dolomite produced by the reaction is highly hydrophilic, and therefore, it needs to be sealed in storage and the sealed-storage time should not be too long. To obtain an economical and effective formulation of raw materials, the ratio of ferrosilicon and fluorite powder used in the preparation of briquettes must be dynamically adjusted according to the composition and morphology of the calcined dolomite. Also, the density and looseness of the prepared pellets must be determined according to the chemical compositions of the dolomite that is extracted from different areas.

The key step in the Pidgeon process is the reduction. The reduction directly affects the length of the production cycle and the quality of the final product. The activity of the calcined dolomite, briquette-making pressure, and formulation of components in the early stage all affect the reduction [45].

To make the reduction proceed more thoroughly and economically, the amount of ferrosilicon and fluorite powder used in the preparation of the briquettes must be dynamically adjusted according to the characteristics of the calcined dolomite, and the briquetting pressure for preparing the briquettes must be reasonably determined according to the chemical composition of the dolomite ores. Through such adjusting and determination, the optimal formulation of raw materials and the parameters can be determined.

The Pidgeon process makes full use of resources such as dolomite, coal, and ferrosilicon in the central and western regions of China. Because of low labor costs, reduction furnaces in foreign countries that use heavy oil and electricity as fuels can be converted to those that use coal as fuel, and this greatly reduces the production costs. Thus, this process has gradually replaced the electrolytic reduction process. To produce 1 t of metallic magnesium, the consumption ratio of dolomite, ferrosilicon, and fluorite as the main raw materials is about 15:1.3:0.21. The consumption of the main primary energy is about 4 t of standard coal when an advanced regenerative reduction furnace is used, and 7–10 t of standard coal is consumed when an old-style reduction furnace directly fired by coal is used. Therefore, this is a high energy consumption industry for primary energy, and the energy consumption per 10,000 yuan of output value is more than 2 t of standard coal. Jiang Hanxiang et al. [48] studied the main factors that affect the recovery rate of magnesium metal during the preparation of magnesium via the Pidgeon process. The results indicate that when the calcination temperature is 1230 °C and the ratio of raw materials is 1:1.15, the formation rate of magnesium reached maximum value after a reduction cycle of 10 h, and the average magnesium recovery rate exceeded 75%.

The advantages of the Pidgeon process include the following: short process, fewer equipment requirements, simple operation, low investment, short plant construction period, and suitable for primary energies such as coal, natural gas, heavy oil, and gas. In the production process, no harmful gases are generated or discharged. By-product slag can also be used as a raw material in the production of cement and fertilizer. Thus, many small and medium-sized enterprises in China use the Pidgeon process to produce magnesium [49]. In addition, dolomite minerals are widely distributed in China; the reserves are rich with remarkable quality, and this provides unique conditions for developing magnesium smelting using the Pidgeon process.

However, in recent years, with the requirements of the national energy-savings and emission-reduction policies and with the awareness of the need for environmental protection, the adverse effects that result from the Pidgeon process have become distinctively obvious [50]. The more severe problems are the huge energy consumption and the generation of serious environmental pollution. To solve both problems, more in-depth research is necessary regarding the thermal decomposition of dolomite and the thermal reduction process of MgO. Combined with the life cycle theory, the current technological level can be modified and improved. The economic and environmental benefits of magnesium production can be comprehensively evaluated, and novel magnesium smelting processes that are green and energy-saving can be developed. A novel green and energy-saving magnesium smelting process can provide technical support for China's magnesium smelting industry using the Pidgeon process to improve the production standard, reduce and save energy consumption, and carry out the automation and mechanization of the magnesium alloy industry.

(2) Balzano method. The Balzano Process evolved from the Pidgeon process and originated in a small magnesium smelting plant in Balzano, which is a town in Italy. At present, the Balzano process is used by Brasmag, Brazil. The Balzano process uses a larger vacuum reduction tank and adopts internal electric heating during vacuum reduction; thus, the energy consumption of this method is much lower than that of other reduction methods. The production raw material is still dolomite. Different from other thermal reduction methods, in the Balzano Process, after the calcined dolomite and ferrosilicon are pressed into briquettes and placed in the reduction tank, an electric heater is used directly to heat the briquettes instead of heating the entire reduction tank. The pressure inside the tank is 3 Pa, and the reaction temperature is 1200 °C. The heating furnace consumes only 7–7.3 kWh to produce 1 kg of magnesium, and other production process parameters are similar to those of the Pidgeon Process [51]. This indicates that the energy consumption of the Balzano Process is significantly lower than that of other thermal reduction methods.

(3) Magnetherm method, also known as the semicontinuous thermal reduction method [52]. This method originated in France and was proposed by Pechiney Aluminum Company around 1960. The Magnetherm process soon became the main method for producing magnesium in the northwestern United States. Different from the Pidgeon process, the steel shell of the sealed reduction furnace that is used in the process is lined with insulation materials and carbon materials so that it has a high reaction temperature (1300–1700 °C) when it is heated internally by electric resistors [53]. The raw materials in the reduction furnace include calcined dolomite, ferrosilicon, and calcined bauxite, which can decrease the melting point of slag. The heat generated by the current passing slag can keep the temperature of the furnace at 1723–1773 K; liquid slag can be

directly extracted, which does not destroy the vacuum in the furnace. Semicontinuous thermal reduction uses continuous feeding and intermittent slagging. It does not generate harmful gases and has a larger production capacity, although the cost is higher [54]. The reduction reaction in an electric furnace is carried out under vacuum conditions. This method uses dolomite and bauxite as raw materials and Si (Fe) as a reducing agent. The main features of the Magnetherm process are similar to those of the Balzano method. They all use electric heating in a reactor. Generally, the temperature in the reactor is in the range of 1300–1700 °C, and all of the materials are liquid in the magnesium smelting furnace that is used in the Magnetherm Process. There are two main reasons why the Magnetherm process requires such a high-temperature calcination: First, when a large amount of raw materials are fed into the reactor, the reactor still needs to maintain a degree of vacuum (0.266–13.3 kPa). Second, the reduction reaction requires high temperature conditions. In the Magnetherm process, magnesium vapor is concentrated on the condenser in a gaseous or liquid state, and the entire production cycle is 16–24 h. The daily magnesium output of the Magnetherm process is generally 3–8 t, and every 7 tonnes of raw materials consumed can produce 1 ton of metal magnesium.

(4) MTMP method. In 2004, Mintek and Eskom in South Africa jointly published the Mintek thermal magnesium process (MTMP), which is also called the South African thermal method because this method was jointly developed by two companies that are in South Africa. MTMP uses ferrosilicon as a reducing agent to extract Mg from dolomite or MgO with an electric arc furnace at a reaction temperature that is controlled to be between 1700 and 1750 °C. Finally, magnesium vapor is enriched in a condensing chamber and takes a liquid form [51]. MTMP allows instantaneous discharge of waste slag, and the reduction process is carried out under atmospheric pressure. When discharging waste slag and extracting magnesium, the vacuum environment does not need to be cut off, and so continuous production can be achieved.

With the continuous development of magnesium smelting technology, MTMP has been constantly optimized. In 2002, the quality of metallic magnesium produced via MTMP can reach more than 80%, but there is still the problem of not taking magnesium in time, which means that the production process cannot work continuously. In October 2004, a new type of condenser was used in MTMP, and the entire production cycle was 8 days. The condensing device in MTMP magnesium smelting includes: a melting furnace, industrial elbow, electric arc furnace, second condensing chamber, stirrer, over-pressure protection device, and piston for cleaning particles.

In MTMP, the temperature of the electric arc furnace is in the range of 1000–1100 °C, the average feed rate of raw materials is 525 kg/h, and the ratios of the feed materials are about 5.5% Al, 10.7% Fe, and 83.8% dolomite; the ratios of the feed materials can be controlled using a valve. The mixed raw materials undergo reduction reactions in the reaction furnace to generate magnesium vapor. The magnesium vapor is condensed into liquid magnesium as it passes through the industrial elbow. The liquid magnesium is enriched in the furnace (the upper end of the furnace is equipped with a secondary enrichment device to enhance the purity of the magnesium), and magnesium can be regularly extracted as the final product through the lower end of the furnace [55].

1.5 Industrial Policy of China's Magnesium Industry

Since 2001, the National Development and Reform Commission and the Ministry of Science and Technology have listed magnesium alloy as a priority industry for development; the Ministry of Science and Technology has listed *Development and Industrialization of Magnesium Alloy Applications* as a major scientific and technological research project during the 10th Five-Year Plan. In the *Interim Provisions on Promoting Industrial Structure Adjustment* and *Industrial Structure Adjustment Guidance Catalog* promulgated by the National Development and Reform Commission on December 21, 2005, the casting of high-quality magnesium alloy and the processing technologies for plates, pipes, and profiles were included in the encouraged development projects. In 2006, the Ministry of Science and Technology promulgated the *Key Technology Development and Application for Magnesium and Magnesium Alloy* as a major support project in "11th Five-Year Plan". The nation invested another 50 million yuan as a guide for funding the development of key technologies and applications of magnesium. The explicit orientation of the national policy is conducive to the transformation of China's magnesium industry from primary magnesium production to the production of high-tech and high-value-added deep-processed products. These can accelerate the transformation of mainland China's magnesium industry from resource advantages to economic advantages, and they can convert China from being the major magnesium producer to being a powerful country in the magnesium alloy industry [56].

In 2009, China carried out industry integration regarding the policies issued by relevant departments. In *Promoting the Adjustment and Optimization and Upgrading Plan of the Raw Material Industry in the Central Region* ([2009] No. 664 of the original Ministry of Industry and Information Technology), the Ministry of Industry and Information Technology required the elimination of "small nonferrous" enterprises that had a magnesium output lower than 10,000 tonnes per year. The *Shanxi Province Metallurgical Industry Adjustment and Revitalization Plan*, which was issued in 2009, clearly requires that the comprehensive energy consumption of magnesium smelting enterprises should be less than 5.6 t of standard coal/t. At the end of 2011, all magnesium producers with an annual output that was lower than 10,000 tonnes would be eliminated. By the end of 2015, all magnesium producers that had an annual output of lower than 20,000 tonnes would be eliminated.

Originally organized by the former National Development and Reform Commission and now drafted by the Department of Raw Material Industry of the Ministry of Industry and Information Technology, the "*Magnesium Industry Access Conditions*" clarifies strict access requirements in seven aspects: the company's layout and scale, process equipment, product quality, resource and energy source consumption, environmental protection, safe production and occupational hazards, and supervision and management. Among these aspects, there is a requirement for the scale of an enterprise, which states, "The annual production capacity of existing enterprises should be greater than 15,000 tonnes; the annual production capacity of renovated and expanded enterprises should be greater than 20,000 tonnes, and the annual production capacity of new magnesium enterprises should be greater than 50,000 tonnes". According to the requirement, the annual energy output of the eliminated enterprises will reach 783,000 tonnes, and the annual production capacity of the remaining enterprises will be only 536,000 tonnes.

At present, a number of magnesium industry producing groups and industrialization bases have begun to take shape in China. They are mainly distributed in eastern areas such as Jiaodong Peninsula and the Yangtze River Delta, in western areas such as Qinghai, Ningxia and Chongqing, in southern areas such as the Pearl River Delta, and in northern areas such as Liaoning, Jilin, Heilongjiang; Also, they are in Henan, Beijing, and other areas. Thus, it is seen that an industrial chain related to innovative technology of magnesium and magnesium alloys has initially formed. The industrial chain runs through China's east, west, north, and south, showing a high-tech pattern that ranges from raw material production, manufacturing of production equipment, and development of magnesium alloy products to the formation of demonstrated industrialization bases, thereby promoting changes in the structure of China's magnesium industry and driving the magnesium industry toward deep processing. Before 2015, there were no large-scale magnesium projects that were put into production abroad, and they will still rely on China's supply. At the same time, a series of factors (such as new materials, a low-carbon economy, and policy guidance) have made domestic companies optimistic about the magnesium industry and driven a large amount of social investment, laying the foundation for China to make the leap from being a country with abundant magnesium resources to being a powerful country in applications of magnesium and magnesium alloys [55].

At present, the magnesium output of Shaanxi, Shanxi, and Ningxia provinces are 355,000 tonnes, 79,000 tonnes, and 60,000 tonnes respectively. Collectively, this accounts for 86.7% of the national output. The magnesium smelting industry has achieved a profit of −40 million yuan, showing a year-on-year decrease of 40 million yuan.

From January to October 2018, China's magnesium output was 570,000 tonnes, which was a decrease of 22.4% compared to the same time period in the previous year. The export volume of various magnesium products was 324,000 tonnes, which was a decrease of 15.4%. The export value was 810 million USD, which was a decrease of 9.3%. Among the magnesium products, the total export volume of magnesium ingots was 163,000 tonnes, which was a decrease of 20.2%. The export volume of magnesium alloys was 89,000 tonnes, which was a decrease of 10.9%, and the export volume of magnesium powder was 63,000 tonnes, which was a decrease of 11%.

In November 2018, the average domestic price of magnesium spot was 18,206 yuan/ton, which was a 27% increase compared to the previous November. From January to November, the average domestic spot price of magnesium was 16,359 yuan/ton, which was a 10.2% increase compared to the same time period in the previous year. Since 2018, the price of magnesium maintained an upward trend with some fluctuations, and it increased to 18,500 yuan/ton in early November, which was close to the highest point in recent years [57].

In 2018, the overall operation of the magnesium industry was steady, domestic consumption increased, and the price of magnesium continued to increase. However, the stress that resulted from environmentally friendly renovation of the smelting process, which was more intense, and the application of deep-processed products needed to be accelerated. Thus, the transformation and upgrading of the magnesium industry remained arduous.

First, the output of primary magnesium decreased year-on-year, and the price went up with fluctuations. According to 2018 statistics from the Nonferrous Metals Association, magnesium output was affected by environmental protection and production restrictions. Specifically, China's output of primary magnesium was 860,000 tonnes with a year-on-year decrease of 5.4%. Contraction on the supply size supported an upward trend in the price of magnesium. The annual average spot price of magnesium was 16,488 yuan/ton, with a year-on-year increase of 10.5%. According to a survey organized by an industry association, the actual profitability of magnesium smelting companies has increased year-on-year, and industry benefits have continued to improve.

Second, the domestic consumption of magnesium continued to grow while the volume of exported magnesium fell. In 2018, China's magnesium consumption was 450,000 tonnes with a year-on-year increase of 7%, and there was a 2% increase in the magnitude. However, because of the decrease in the consumption of 3C products, growth in the consumption of processed magnesium material decelerated. The demand for magnesium from abroad has fallen, and the total volume of various exported magnesium products in the entire year was 410,000 tonnes with a year-on-year decrease of 11%. The volume of exported magnesium products accounted for 48% of the magnesium output.

Third, industrial structures were adjusted, and there were breakthroughs in high-end applications. In 2018, Baowu Iron and Steel Group invested in Yunhai Metal to jointly expand the applications of magnesium, and they achieved a strong collaboration between state-owned and private enterprises. Henan Hebi Magnesium Trading Center was established to promote the circulation of magnesium products and to promote healthy operation of the domestic magnesium market via innovative trading modes. New progress has also been made in the production and application of high-end magnesium materials. For instance, automobile wheels that consist of forged magnesium alloys have been industrialized and exported to European and American markets. Large-scale and stable preparation of new controllable and degradable magnesium alloy materials have been achieved for bone repair and have been supplied to domestic medical device companies.

Fourth, environmentally friendly renovation is under heavy stress, and the task of green development is arduous. In recent years, domestic magnesium smelting technology has undergone continuous modification while the levels of mechanization and automation have remained relatively low. Working conditions need to be improved, and the level of energy savings, emission reductions, and waste slag recycling urgently need improvements. In 2018, with in-depth advances in pollution prevention and control, some magnesium smelting enterprises terminated production because of the renovation requirements that are based on environmental protections. The difficulty of employment of enterprises became prominent, and the task of achieving green development of the magnesium industry has been arduous.

In 2019, the development of lightweight transportation at home and abroad provided more opportunities for expanding magnesium applications. Meanwhile, stricter requirements were put forward for improving the level of green development for the entire magnesium industry chain. Modifying smelting technology and expanding magnesium applications will be important for promoting high-quality development of the magnesium industry. The Ministry of Industry and Information Technology will continue to promote relevant local governments and enterprises to establish platforms for research and development of magnesium smelting technology. They also support the magnesium industry in implementing technological transformations for green production, encourage wider application of key products such as magnesium wheels, and accelerate large-scale application of the magnesium industry [58]

1.6 Problems with Magnesium Slag After Smelting

1.6.1 Properties and Damage of Magnesium Slag

Two major issues that China faces are energy and the environment. In 2014, the annual magnesium output of China reached 873,800 tonnes [59]. However, for every tonne of metallic magnesium that was produced via smelting, approximately 6.5–7 tonnes of magnesium slag were produced. Therefore, based on the current output, more than 6 million tonnes of magnesium slag was produced in China last year. Considering the cumulative effect of the past two decades, the total amount of magnesium slag is huge.

Magnesium slag is an industrial solid waste that is produced in the process of preparing magnesium metal via electrolysis or thermal reduction with magnesium-rich minerals, such as dolomite, serpentine, and magnesite [60]. Magnesium slag is a gray alkaline block or powder. After slag absorbs moisture, a suspension that is very alkaline (pH around 12) and has stable properties forms. As a result, the soil ever stacked with magnesium slag is easy to harden and salinize; this endangers the growth of crops and affects the future normal use of land. The disordered and random accumulation of a large amount of magnesium slag enters rivers and groundwater systems with rainwater leaching and infiltration; this can change the pH of a water body and seriously affect the ecological security of water resources.

From the perspective of chemical composition, magnesium slag is mainly composed of CaO, SiO_2, Al_2O_3, MgO, and Fe_2O_3. Because of different sources of magnesium ores and different production processes of smelted magnesium, the content of each component is not constant and exhibits a certain amount of fluctuation.

Magnesium slag is a by-product of magnesium smelting. With the rapid growth of magnesium production, the harm of magnesium slag has been obvious. Thus far, magnesium slag cannot be effectively used and the only treatment is dumping and filling. Furthermore, magnesium slag is easily weathered into powder-like matters in the wind; it does not easily settle out of the wind, and the polluted surfaces are difficult to control.

The harm of magnesium slag to the human body is manifested in various ways. Long-term inhalation of concentrated dust can cause diffuse and progressive fibrosis-based systemic diseases (pneumoconiosis). Magnesium dust can be dissolved on the bronchial wall, where it can be absorbed and then carried by the blood to the whole body, resulting in systemic poisoning. Contact or inhalation of dust can cause local irritation to the skin, cornea, and mucosa, producing a series of pathological changes.

The harm that magnesium slag has on the environment is mainly manifested in the following two ways:

(1) Magnesium slag easily causes dust pollution. After reduction, magnesium-containing slag is dust that has a high content of fine powder, and 60–70% of the particles are less than 160 μm in diameter. After reduction, the completely pulverized magnesium slag can basically pass through a 200-mesh sieve, and its fineness is similar to that of cement. These magnesium slag particles are finer than common coal ash dust. They can be suspended in the air and do not settle easily; thus, they easily form dust pollution in the natural environment. Furthermore, they are easily inhaled into the respiratory tract and can cause respiratory diseases. In addition, the dry and windy climate of the northern region exacerbates the severity of damage.

(2) Magnesium slag easily causes soil compaction. After reduction, magnesium slag has a strong ability to absorb moisture, and this can easily cause soil salinization and compaction of soil. The land and surrounding areas on which magnesium slag has accumulated basically cannot be used for agricultural cultivation any more, and this results in the reduction of a large amount of farmland resources. With a reduction in China's arable land and an increase in global food prices in recent years, the harm of magnesium slag on cropland has gradually become severe [61].

1.6.2 Treatment and Efficient Utilization of Magnesium Slag

Magnesium smelting industry is an industry that consumes a high amount of energy and materials; meanwhile the pollution that it generates in production is severe. Although cleaner coke oven gas has been used to replace coal, the by-product of magnesium smelting (reducing slag) has not been well treated. With a continuous increase in magnesium output, the environmental pollution caused by magnesium slag is attracting increasing attention. How to combine our own characteristics and local conditions to reasonably and effectively utilize magnesium slag is a difficult problem that we face.

Although China has many favorable conditions for developing the magnesium industry, in the increasingly intense market competition, the magnesium industry still faces several problems: relatively old technical equipment, low thermal efficiency, high labor intensity, and serious pollution, especially the large amount of waste slag that is produced in the production process. In the 2010 edition of *"Industrial Structure Adjustment Guidance Catalog"* promulgated by the National Development and Reform Commission, magnesium smelting projects are still listed in the restricted category. This categorization is conducive to curbing the low-level construction and disorderly development of the magnesium industry; it can also promote the optimization of each step in the Pidgeon process and the comprehensive use of waste residue from magnesium smelting as a resource.

According to the *"State of the Environment Bulletin of China"* [62], which was published by the Ministry of Environmental Protection of the People's Republic of China, the national discharge of industrial solid waste gradually increased during the period of 2000–2014. In the past five years, the annual growth rate of industrial solid waste reached 10%. The solid waste that is generated by industries such as electric power, heat production and supply, metal smelting and processing, and mining of nonferrous metals accounted for about 80% of all solid wastes [63]. On the whole, China's industrial solid waste is still in a stage where there is large production and a low utilization rate.

At present, there is no effective treatment method for magnesium slag; thus, it is mainly treated via stacking and burying methods like landfill in mountain depressions and dumping in wasteland, and this results in a very low utilization rate. The massive discharge and accumulation of magnesium slag requires a lot of land resources [64], causes dust pollution, and has direct impacts on crops and the surrounding environment. Magnesium slag hardens the land so that arable land that is polluted with magnesium slag loses its fertility. This leads to decreased crop yields and reduced area of arable land. Magnesium slag flows into rivers with rainwater, has a great impact on bodies of water, and seriously endangers human health and the growth of crops. Therefore, magnesium slag has become a major hazard that results from the magnesium industry. With the rapid development of the magnesium industry, the amount of magnesium slag is increasing, and the harm it causes to the environment is attracting more attention. Thus, it is clear that the harm caused by magnesium slag is urgently needed to deal with. It is particularly crucial to find a scientific and reasonable way to solve the pollution of magnesium slag, and the approach should bring huge social and ecological benefits [61].

China is a major magnesium-producing country and has relatively less applications of magnesium slag then developed countries. China has abundant magnesium resource storage and a large export volume, and there are fewer magnesium producers abroad. There are very few studies on applications of magnesium slag that have ever considered it as an industrial waste. Thankfully, in response to the problem of using magnesium slag as a resource, domestic researchers have carried out a lot of research and made certain achievements. Some magnesium plants in China have applied magnesium slag in road paving, brick production, and cement production for use in construction. However, most magnesium slag is still disposed on in landfills. In view of the potential value of magnesium slag, domestic researchers have carried out research work in related applications and have obtained some research results; the results have mainly focused on the use of magnesium slag in preparing cement clinker or admixtures, building bricks, wall materials, cementitious materials, etc.

Magnesium slag is rich in CaO, SiO_2, MgO, and other components that have higher alkaline oxide contents [65]. In the production of cement, magnesium slag can be used as alkaline substances that can be added to raw materials, thereby improving the quality of cement and enhancing its stability [66]. In the production of wall materials, introducing finely-ground magnesium slag powder can promote various chemical reactions between components in the raw materials and can stimulate the activity of magnesium slag; thus, low-cost, high-strength, and low-density wall materials can be produced through a simple and easy process [67]. Magnesium slag can also be used as a desulfurizer because the amount of calcium oxide in magnesium slag is relatively high. Substituting magnesium slag for some of the calcium oxide in a circulating fluidized bed boiler can achieve the desulfurization effect [68].

Results of heavy metal leaching toxicity for magnesium slag are shown in Table 1.19 [57]. Regardless of whether the HJ method with water leaching or TCLP was leaching with a buffer solution (pH = 2.88) was used, the leaching results were

Table 1.19 Heavy metal leaching toxicity results for magnesium slag

| Heavy metal | HJ557-2010 leaching test | | | | TCLP leaching test | | | NEN7341 |
	Detected value (mg/L)	Limit(mg/L)	mg/kg		Detected (mg/L)	Limit (mg/L)	mg/kg	mg/kg
Cr	0.012 ± 0.002	10	0.12 ± 0.02		0.034 ± 0.003	5	0.68 ± 0.06	0.75 ± 0.11
Cu	0.043 ± 0.004	50	0.43 ± 0.04		0.021 ± 0.005	15	0.42 ± 0.10	2.34 ± 0.021
Zn	<0.001	50	<0.01		<0.001	25	<0.02	4.52 ± 0.83
Ni	<0.01	10	<0.01		0.008 ± 0.002	20	0.16 ± 0.04	3.20 ± 0.55
Pb	Not detected	3	<0.1		<0.01	5	<0.2	<0.50
Cr^{6+}	Not detected	1.5	Not detected		Not detected	–	Not detected	Not detected
Cd	Not detected	0.3	Not detected		Not detected	1	Not detected	Not detected
As	Not detected	1.5	Not detected		Not detected	5	Not detected	Not detected
Hg	Not detected	0.05	Not detected		Not detected	0.2	Not detected	Not detected

far below the standard limit, and this indicates that magnesium slag is not a hazardous waste containing heavy metals and can be used in agricultural resource utilization.

Scientific and technological researchers have conducted a lot of research on the resource utilization and harmless use of magnesium slag. Zhu Guangdong et al. [65] added reducing slag into a three-wire electric furnace, heated the reducing slag to a molten state, then turned water-quenched slag into cement raw material after a series of operations, such as water quenching. This approach is an effective and feasible method for solving the problem solid waste discharge in the magnesium industry. Han Tao et al. [69] invented a method for preparing high-performance magnesium slag. By spraying a dilute acid solution on high-temperature magnesium slag when it is just removed from the tank, the expansion rate of the magnesium slag was reduced while the content of active ingredients in the mineral increased. In recent years, environmental pollution problems such as stacking of magnesium slag have gradually become burning question. Li Dongxu et al. [70] invented a method for making magnesium slag bricks without using sintering process. By adding 50–70% magnesium, slag bricks made using this method have a compressive strength that reaches 35.6 Mpa in 38 days. The nonsintered bricks have a series of characteristics, such as high strength grade, early strength, low shrinkage, and good frost resistance. Han Tao et al. [71] aimed at the slow hydration speed and slow strength growth of C_2S (dicalcium silicate) in magnesium slag raw materials and provided a method for preparing nonsintered bricks using magnesium slag. The added amount of reduced magnesium slag was 30–50%, and the 28-day strength was 19.7–36.8 Mpa. Wu Laner et al. [72] invented a phosphorus compound that can stabilize β-dicalcium orthosilicate in reduction slag generated from magnesium smelting via the Pidgeon process. The modified magnesium slag can be used as a concrete admixture or cement admixture. Wu Laner et al. [73] also invented a magnesium slag modifier and a modification method for magnesium slag. In this method, β-dicalcium orthosilicate can be converted to γ-dicalcium silicate during the Pidgeon process. With the addition of a small amount of boric acid and the use of a high-temperature roasting process, the problem of pulverizing magnesium slag is solved. Wu Yong [74] crushed magnesium smelting slag into power and then dissolved it with acid and a precipitation agent to make $CaSO_4 \cdot 2H_2O$ precipitate from $CaCl_2$ solution. The filtered $CaSO_4 \cdot 2H_2O$ precipitate can be produced into gypsum powder, and the filtrate can be used as liquid nitrogen fertilizer. Therefore, the comprehensive treatment of magnesium slag is the main subject of the sustainable development of the magnesium industry.

References

1. Kedan J (2016) study on hydration reaction kinetics of quenching magnesium slag. Taiyuan University of Technology, Taiyuan
2. Mordike BL, Eber T (2001) Magnesium properties application potential. Mater Sci Eng A 302:37–45
3. Yuanyuan Z (2013) Researching on the dynamics and parameters' optimization of smelting magnesium by silicon-thermo-reduction. Jilin University, Changchun

 4. Zhiguo L, Shundu C (2003) Deep processing and application of dolomite in chemical industry. Chem Mineral Process (1):4–7
 5. Yiming L, Xinping Y (2003) Comprehensive development and utilization of dolomite. Yunnan Chem Ind 30(1):5–7
 6. Zhiguo L, Shundu C, Jiandong Z (2003) Exploitation and application of the series of products for dolomite. Mineral Compr Utilization (2):27–33
 7. Liping G (2008) Bright prospect of magnesium industry. Aluminum Process 05:48–52
 8. Chongyu Y (1991) Metallurgy of lightweight metals. Beijing Metallurgical Industry Press, pp 225–226
 9. Gregg JM, Sibley DF (1984) Epigenetic dolomitization and the origin of xenotopic dolomite texture. J Sediment Petrol 54:908–931
10. Sibley DF, Gregg JM (1987) Classification of dolomite rock textures. J Sediment Petrol 57(6):967–975
11. Cong Y, Huizhong Z, Han Z et al (2012) Influence of technological parameters on activity of light-burnt dolomite. Refractories 46(1):48–50
12. Limei B, Yuexin H (2005) Study on the preparation of high-quality active magnesium from magnesite. China Non-ferrous Min Metall 7(21):47–48
13. Yue Q (2002) How does magnesite industry meet the challenge of China's entry into WTO. Land Resour 1:22–23
14. Zhaoyou C, Hongxia L (2005) Comprehensive utilization of natural magnesium-containing resources and development of MgO-based refractories. Refractories 39(1):6–15
15. Sumei D (2001) Magnesite resources and market in China. Nonmetallic ore 24(1):5–6
16. Riyao X (1993) Magnesium metallurgy, Revised ed. Beijing Metallurgical Industry Press, p 11
17. Baojun Y, Shaoming Y, Chengxiang S (2003) A new technology of comprehensive utilization of serpentine. Min Metall Eng 23(1):47–49
18. Pu W (2005) Mineral resources of serpentine fibre and its environment in China. China Non-metallic Mineral Ind Guide (1):50–52
19. Peihua M (2000) Comprehensive utilization of salt lake resources. Prog Geosci 15(4):365–375
20. Chun W (2010) Preparation of calcium perborate and magnesium peroxide. Dalian University of technology, Dalian
21. Xiping H, Qi Z, Shuyuan G, Gongwei W (2004) The present state and future outlook of the exploitation and utilization of magnesium resource in sea water and brine about our country. Haihu Salt Chem Ind 33(6):1–6
22. Qi J, Hong J (2005) China's mining industry has a long way to go. Chinese J Non-metallic Min 2(46):59–61
23. Yufei W, Yifeng C (2000) Analysis on utilization of magnesium resources in Chaerhan Salt Lake. World Nonferrous Metals 12:14–15
24. Hoy-Petersen N (1990) In: Proceedings 47th annual world magnesium conference. International Magnesium Association
25. Pidgeon LM, Alexander WA (1944) Thermal production of magnesium-pilot plant studies on the retort ferrosilicon process. Trans AIME 159:315–352
26. Osborne R, Cole G, Cox B, et al (2000) USCAR project on magnesium structural casting. In: IMA-proceeding. international magnesium association, pp 1–5
27. Hongying S (2004) Prospect of world magnesium industry production and technology. World Nonferrous Metals 8:36–39
28. Beals RS, Tissington C, Zhang X et al (2007) Magnesium global development: Outcomes from the TMS 2007 annual meeting. JOM 59(8):39–42
29. Tongjun Z, Xingguo L (2002) Application of magnesium alloy and magnesium industry in China. Mater Guide 16(7):11
30. Hou Y (2017) Study on magnesium slag desulfurizer modified by additives in quenching hydration. Taiyuan University of Technology, Taiyuan
31. Jing Y, Yaowu W (2019) Reduction mechanism of magnesium smelting by Pijiang process. J Process Eng 19(3):561–566

32. Hongxiang L, Yongnian D, Wenhui M, Yifu L (2007) The research direction of China's magnesium industry in the future. Light metals (1):46–49
33. Brooks G, Trang S, Witt P, Khan MNH, Nagle M (2006) The carbothermic route to magnesium. JOM 58(5):51–55
34. Jianping P, Shidong C, Xiaolei W et al (2009) Study on thermal reduction of calcined dolomite by calcium carbide. Light Metals 3:47–49
35. Yang J, Kuwabara M, Liu ZZ, Asano T, Sano M (2006) In situ observation of aluminothermic reduction of MgO with high temperature optical microscope. Iron Steel Inst Japan 46:202–209
36. Byron B (1997) Global overview of automotive magnesium requirements and supply and demand. International Magnesium Association (IMA)
37. Riyao X (2003) Magnesium production technology. Central South University Press, Changsha, p 12
38. Vecchiattini R (2002) DEUIM. Ph D. Thesis. University of Genoa, Italy
39. Carlson KD, Margrave JL (1967) The characterization of high-temperature vapors. In: Margrave JL (ed). Wiley, NY
40. Stanley RW, Berube M, Celik C et al (1996) The Magnola process magnesium production. In: IMA 53. Magnesium—a material advancing to the 21 st Century, pp 58–65
41. Sharma RA (1996) A new electrolytic magnesium production process. JOM 48(10):39–43
42. Duhaime P, Mercille P, Pineau M (2002) Electrolytic process technologies for the production of primary magnesium. Mineral Process Extr Metall 111(2):53–55
43. Faure C, Marchal J (1964) Magnesium by the magnetherm process. J Metals 16(9):721–723
44. Liping Duan (2015) Study on fractal characteristics of hydration products from quenching magnesium slag. Taiyuan University of Technology, Taiyuan
45. Riyao X (2003) Production technology of silicothermic magnesium smelting. Central South University Press, Changsha
46. Yuesheng C, Gang S, Aisheng L (2005) Knowledge of magnesium and magnesium alloy production. Metallurgical Industry Press, Beijing
47. Basu P, Sarkar SB, Ray HS (1989) Isothermal reduction of coal mixed iron oxide pellets. Trans Indian Inst Metals 42(42):165–72
48. Hanxiang J, Qiqiang Z, Hong G (2006) Study on producing magnesium technology for dolomite. J Chongqing Univ (Natural Science Edition) 29(9):64–67
49. Ding W, Zang J (2001) The Pidgeon process in China. In: 3rd annual Australasian magnesium conference, Sydney, Australia, p 7
50. Ramakrishnan S, Koltun P (2004) Global warming impact of the magnesium produced in China using the Pidgeon process. Resour Conserv Recycl 42(1):49–64
51. Jilong H, Qingguo S (200) Progress of magnesium production process. Salt Lake Res 4:59–65
52. Goswami MC, Prakash S, Sarkar SB (1999) Kineties of smelting reducting of fluxed composite iron and pellets. Steel Res 70(2):41–46
53. Weixue Y, Zaihua F, Tongfei L, Guanghua C (2009) Comprehensive utilization of coke oven gas review. Hebei Chem Ind 12:34–36
54. Wilson CB (1996) Modern production processes for magnesium. Light Metals 1996:1075–1079
55. Chong W (2013) A study on the dolomite calcination process and a new technology of product magnesium thermal reduction process based on the LCA theory. Jilin University, Changchun
56. Xiangbin X (2011) Experimental study on magnesium smelting slag used as sulfur fixing agent for coal combustion. Ganzhou, Jiangxi
57. Ministry of industry and information technology. http://www.miit.gov.cn/n1146312/n1146904/n1648356/n1648358/c6533590/content.html
58. Ministry of industry and information technology. http://www.miit.gov.cn/n1146312/n1146904/n1648356/n1648358/c6666287/content.html
59. Magnesium industry branch of China nonferrous metals industry association. Statistics of raw magnesium production in China from January to December 2014 [EB/OL]. http://wvyw.chinamagnesium.org/tjsj.php?catPath=0.1.133.135, 2015-02-01
60. Xiao LG, Wang SY, Luo F (2008) Status research and applications of magnesium slag. J Jilin Inst Archit Civil

61. Zhengguan L (2012) Research on the technology and material related properties of magnesium slag as raw material for making grate brick. Taiyuan University of Science and Technology, Taiyuan
62. Ministry of environmental protection of the people's Republic of China, environmental statistics bulletin. http://www.mep.gov.cn/zwgk/hjtj/
63. Ningning W, Chen W (2014) Research on comprehensive utilization technology of industrial solid waste resources. Agric Technol 34(2):251–252
64. Yuming T, Shaopeng Z, Kaiyue W, Zhankao C, Yuesheng C (2014) New research of magnesium slag reutilization. Shanxi Metall 1:1–4
65. Cai JW, Gao GL, Bai RY et al (2011) Research on slaked magnesium slag as a mineral admixture for concrete. Adv Mater Res 287–290:922–925
66. Thomas JJ, Jennings HM, Chen JJ (2009) Influence of nucleation seeding on the hydration mechanisms of tricalcium silicate and cement. J Phys Chem C 113(11):4327–4334
67. Enqing C, Lianping W (2006) Research on production technology of blown-out concrete made from magnesium-reduced slag and fly ash. J Three Gorges Univ (Natural Science Edition) 25(6):522–523
68. Xiaolei Qiao, Yan Jin (2007) Experimental study on desulfurization performance of reduction slag from magnesium smelting. Sci Technol Inform Dev Econ 17(7):185–187
69. Tao H, Xiuzhi J, Huiqi W, Xueteng J, Fang Y (2015) A preparation method of high-performance magnesium slag. CN 104370482, Publication date: 2015.02.25
70. Dongxu L, Chunhua F (2012) A kind of non-fired magnesia slag brick and its preparation method. CN 102503313 a, Announcement date: 20 June 2012
71. Tao H, Fengling Y, Fangqin C, Yanxia G, Fangbin X, Yongsheng L (2013) Magnesium slag unburned brick and its preparation method. CN 102942347 a, Announcement date: 2013.02.27
72. Laner W, Fenglan H, Qixing Y, Chun D, Shengwei G, Yong J, Yuhong C, Youjun L (2012) A method for stabilizing β-dicalcium orthosilicate in magnesium smelting slag by PI Jiang process. CN 102730706 a, Published on 17 October 2012
73. Laner W, Fenglan H, Qixing Y, Yuhong C, Youjun L, Chun D (2011) A magnesium slag modifier and its modification method. CN 102071327 a, Published on: 25 May 2011
74. Yong W (2014) Comprehensive utilization method of slag from magnesium smelting by Pijiang process. CN 103553109a, Published on: 5 February 2014

Chapter 2
Magnesium Smelting via the Pidgeon Process

Abstract At present, there are two main magnesium smelting methods: electrolytic method and reduction method. Compared with the electrolytic magnesium smelting process, the Pidgeon process has the advantages of simple and mature process flow, small investment, short construction period, multiple heat sources, good magnesium quality, low power consumption, and good utilization of coal, natural gas, heavy oil, gas and so on.

Keywords Pidgeon process · Magnesium smelting · Technological process · Facility

China is the world's largest producer of primary magnesium and has a magnesium smelting industry that is mainly based on the Pidgeon process. Magnesium smelting via the Pidgeon process is a process in which raw material (dolomite) is fed into a reduction tank that is externally heated by a reduction furnace, and then the calcined dolomite is thermally reduced to metallic magnesium using 75% ferrosilicon as a reducing agent in vacuum. The method is named after the inventor, L.M. Pidgeon, and has been applied for a long time; thus, it has been considered to be a classic silicothermic method for smelting magnesium. The Pidgeon process has a history of more than 70 years since its birth, and it has been used in industry in China for more than 40 years.

At present, there are two main magnesium smelting methods: electrolytic smelting and reduction smelting. Compared to electrolytic smelting, reduction smelting has the advantages of having a simple and mature process, requiring a small investment, short construction period, can use various heat sources, produces good quality magnesium, consumes a low amount of power, and can use energy sources such as coal, natural gas, heavy oil, and gas. In addition, China has the world's largest reserves of dolomite, which can be mined for more than one thousand years. This creates good resource conditions for the sustainable development of the Pidgeon process. However, the Pidgeon process also has problems, such as using a small sized reduction tank, having a low filling amount of a single tank, low thermal efficiency, and intermittent production, requiring high labor intensity, consuming high

L. Wu et al., *Comprehensive Utilization of Magnesium Slag by Pidgeon Process*,
SpringerBriefs in Materials, https://doi.org/10.1007/978-981-16-2171-0_2

energy, and causing serious environmental pollution [1, 2]. The intermittent operation, low production capacity of a single tank, high energy consumption, and other issues limit the development of magnesium smelting via the Pidgeon process.

2.1 Processes in Magnesium Smelting via the Pidgeon Process

2.1.1 Introduction

Processes in magnesium smelting via the Pidgeon process include dolomite calcination, grinding and pelleting, and vacuum thermal reduction. The equipment that is mainly used for dolomite calcination is the rotary kiln. In a rotary kiln, calcined dolomite is the raw material, ferrosilicon is the reducing agent, and fluorite is a catalyst; these materials are weighed and mixed. After grinding, the mixture is pressed into spheres that are called pellets. The steps of grinding, batching, pelleting of calcined dolomite, and placing of the pellets into a furnace should occur as fast as possible (generally, in no more than 4 h). The pellets should be placed in sealed paper bags to avoid moisture absorption, which can reduce the activity of calcined dolomite and then reduce the yield of magnesium [3]. After that, the pellets are placed into a reduction tank and heated to 1200 °C. The inside of the furnace is evacuated to a vacuum level of 13.3 Pa or higher, and then magnesium vapor is produced. Magnesium vapor forms crystalline magnesium, which is also known as crude magnesium, in a condenser at the front end of the reduction tank. After the crude magnesium is refined via flux, commercial magnesium ingot is produced; commercial magnesium is refined magnesium.

The Pidgeon process uses silicothermic reduction to smelt metallic magnesium. There are two main chemical processes that occur in succession:

(1)   Calcination of dolomite:

$$MgCO_3 \cdot CaCO_3 \xrightarrow{1100-1200\,°C} MgO \cdot CaO(\text{calcined dolomite}) + 2CO_2 \uparrow$$

(2)   Reduction:

$$2MgO \cdot CaO + Si(Fe) \xrightarrow{1200\pm10\,°C} 2Mg + 2CaO \cdot SiO_2$$

The main technological process and basic route for the production of magnesium and magnesium alloys via the Pidgeon process have the same principles at home and abroad. The basic flow chart is as shown in the figure above. Differences in technological advancements and clean production levels of enterprises mainly include the levels of mechanization and automation, variations in fuel and energy consumption, and variations in the equipment that is used in each section. The current status of domestic and foreign magnesium smelting enterprises in these aspects is summarized as follows:

(1) There are three main methods to enhance the levels of mechanization and automation of the Pidgeon process. One method uses a microcomputer-controlled closed joint line consisting of batching, grinding, and pelleting. The second method uses mechanical equipment to accomplish slag-removal, tank cleaning, feeding, and magnesium extraction. The third method uses mechanized continuous casting. Advanced foreign enterprises can achieve all three methods, whereas the current advanced domestic enterprises only achieve two methods. Most magnesium plants in China cannot achieve one method, even achieve none at all. This means that the mechanization levels of these enterprises are still quite low, and corresponding improvement is urgently needed.

(2) Most magnesium plants in China use coal as fuel. The production of every 1 t of magnesium consumes 8–12 t of high-quality coal. In addition to wasting energy source and resources, this also emits a large amount of waste gas, causing serious environmental pollution. As seen from the advanced magnesium smelting enterprises at home and abroad that use clean energy, using clean energy is an effective way to enhance the level of magnesium smelting via the Pidgeon process. At present, some domestic manufacturers use coal-water slurry to greatly reduce the emission concentration of sulfur dioxide and nitrogen oxides that form during burning and to make the smoke and dust emissions reach Ringelmann black degree 0, which greatly improves the environmental impact of magnesium smelting via the Pidgeon process. Weienke Technology Company achieved good results with the use of coal-water slurry as a fuel in the production of magnesium smelting. For instance, the combustion efficiency of coal reached more than 98%, and the expected savings in coal would be approximate 20–30% [4].

(3) At present, in the order of advancing levels, the main calcination kilns (furnaces) that are used at home and abroad are sleeve-type shaft kilns, rotary kilns, and vertical kilns. The rotary kiln is the most widely used equipment for calcining dolomite. Regardless of the structure of dolomite, as long as the process conditions (reasonable kiln speed and feeding quantity) are carefully controlled, dolomite can be calcined to obtain good quality calcined dolomite. The utilization rate of raw material in a rotary kiln can reach over 90%; this is twice high as that of the shaft kiln, which has a utilization rate only 40–50% [5]. At present, it is common in Japan to use a rotary kiln in silicothermic magnesium smelting to calcine dolomite. Japan Tazawa Industry Co., Ltd., has two φ 2.5 m × 55 m rotary kilns for calcining dolomite that has high iron content, and the production capacity is 30,000 t/a [6]. Although a rotary kiln has good technical indicators, it is not the most advanced kiln type because of its high operating cost and low heat utilization rate. In recent years, the sleeve-type shaft kiln, which integrates energy-savings, environmental protection, and better performance, has been extensively used in Europe and Japan.

Practice indicates that the sleeve-type shaft kiln has simple equipment, convenient operation and maintenance, a better working environment, and excellent product quality; thus, it is a new type of kiln that has promising development prospects [7].

(4) The reduction process is the core of magnesium production, and the type of reduction furnace that is used greatly affects the advanced level of the process. A traditional magnesium reduction furnace has a long feed cycle, low efficiency, uneven feeding, and gaps in the tank's upper part; these characteristics result in unsatisfactory heat transfer. In addition, most traditional magnesium reduction furnaces lack a recovery device for waste heat; thus, high-temperature flue gas has been directly discharged into the air, causing serious waste in energy sources and serious damage to the environment [8]. The new dual-regenerative reduction furnace uses dual-regenerative combustion in air and with coal gas as fuel. The reduction furnace uses end-heating along the length of the furnace, and each heating point uses a semi-built-in heat storage burner; coal gas uses a two-position three-way reverse valve, and air uses a two-position four-way reverse valve to accomplish the reverse in direction.

The latest reduction technology in international magnesium smelting is vacuum smelting of metallic magnesium with internal electric heating. This technology heat reaction material from the inside part uses electricity as secondary energy. This can increase the reduction temperature, accelerate the reaction process, shorten the reduction cycle, increase thermal efficiency, and reduce energy consumption while avoiding pollution problems that are caused by the combustion of fuel. At the same time, the modified smelting method can increase the furnace capacity and increase the output of a single furnace. This process does not require expensive metal reduction tanks, and thus, the cost is less. In addition, the technology can be designed as a continuous production with automatic feeding and automatic discharging, both automatic feeding and automatic discharging cause this reduction of labor intensity [1, 9].

2.1.2 Raw Materials in Magnesium Smelting via the Pidgeon Process

The Pidgeon process uses dolomite as the main raw material, ferrosilicon as a reducing agent, and fluorite as a catalyst. After calculating, weighing, and batching, the reduction reaction of the raw material is completed in a reduction tank (furnace) to obtain crude magnesium. The ignition loss of dolomite in calcination is about 47%; this means 47% of dolomite decomposes into CO_2 gas, which is discharged into the air, not including the CO_2 that is produced during the heating of dolomite.

Dolomite is crushed in the mine to meet the required particle size, and it is stored in a plant's dolomite yard. Ferrosilicon and fluorite are directly stored in a plant's warehouse. All reserves are based on the amounts of materials that are required for production in half of a month. In silicothermic magnesium smelting, the calcination effect of dolomite directly affects the reduction yield of magnesium. Different types of kilns, different fuels, and different structural types of dolomite require different calcination conditions and result in different calcination effects. Because of different impurities and structures in dolomite, samples have different physical and chemical properties. At present, dolomite can be roughly divided into two categories: one is amorphous dolomite, which has a network structure; the other is a form of dolomite that has a hexagonal rhombus structure.

1. Amorphous dolomite with a network structure can retain the structural characteristics of dolomite after calcination. The particles of such calcined dolomite have large surface defects, and there are large cohesive forces between particles. Also, the amorphous dolomites are prone to deform under mechanical actions. The activity of amorphous dolomite in reduction is greater than that of dolomite that has a hexagonal rhombus structure.
2. Dolomite that has a hexagonal rhombus structure breaks easily, and it is easier to grind than dolomite that has a network structure, which is already finely ground. Also, dolomite that has a hexagonal rhombus structure does not preferentially adhere to the mill and its reactivity is lower.
3. The difference between the hydration activities of the two kinds of calcined dolomite is slight, but their activities in reduction are very different. These observations indicate that the hydration activity of calcined dolomite only represents its hygroscopicity and does not fully represent its activity in reduction. Generally, when hydration activity is higher, the activity in reduction is higher. However, for the same hydration activity, the activity in reduction is not necessarily the same because of the different dolomite structures.
4. Amorphous dolomite that has a network structure is selected. Because of its high strength and high hardness, the lumpiness of dolomite is easily ensured during crushing, that is, a low crushing rate is ensured.
5. Because of its low lattice energy, amorphous dolomite that has a network structure has a lower heat absorption capacity during thermal decomposition than dolomite that has a hexagonal rhombus structure. That is, the energy that is required for calcination is also lower, and there is less thermal cracking during calcination.

As seen from the above characteristics, dolomite that has a hexagonal rhombus structure after calcination has a high crushing rate and is easy to grind. Thus, it is not suitable for use in a shaft kiln or a mixed-load vertical kiln and is more suitable for use in a boiling furnace or a rotary kiln. The reason for this is that the calcined materials in shaft and mixed-load vertical kilns usually accumulate at the bottom discharge ports. Under the pressure of the upper material layer, calcined dolomite can easily become material that has a smaller particle size. This causes poorer air permeability, and the kiln can easily become blocked easily. This can even affect the normal operation of the entire kiln. However, this problem does not appear in rotary kilns and boiling furnaces.

Amorphous dolomite that has a network structure has high strength, high hardness, and low lattice energy. Thus, it has lower heat absorption during thermal decomposition than dolomite that has a hexagonal rhombus structure. Therefore, ideal calcination effects can also be achieved when such dolomite is calcined in shaft kilns and mixed-load vertical kilns. Such dolomite also exhibits good calcination effects in a rotary kiln, and this indicates that it is suitable for a wide range of furnace types. In summary, even though the same shaft kiln is used in current manufacturers, the fact that calcined dolomite has different activities in reduction is also a normal phenomenon because the structural characteristics of dolomite ore are different.

2.1.3 Processes in Magnesium Smelting via the Pidgeon Process

There are four main processes in smelting magnesium via the Pidgeon process: calcination, pelleting, vacuum thermal reduction, and refining.

(1) Calcination of dolomite: Dolomite is heated to 1100–1200 °C in a rotary furnace (rotary kiln) or shaft kiln to form calcined dolomite ($MgO \cdot CaO$). The specific process is as follows: After the calcined dolomite is crushed and screened, dolomite that has a particle size range of 10–40 mm is moved to the top silo of a vertical preheater using a large inclination belt conveyor. The materials in the silo are fed through a feeding pipe into the preheating box of the preheater. When the materials move slowly down into the preheating box, the dolomite is preheated to about 900 °C by the kiln's tail hot gas, which is at 1000–1100 °C. The rotary kiln provides a high-temperature heat source for the kiln through a burner that is installed in the kiln. High-temperature calcination causes dolomite to undergo thermal decomposition from $CaCO_3 \cdot MgCO_3$ (Scheme 1.2) at 1150–1200 °C to generate MgO and CaO, which are required for the reduction reaction.

$$CaCO_3 \cdot MgCO_3 \rightarrow MgO + CaO + 2CO_2 \uparrow \qquad (1.2)$$

The calcined dolomite is cooled in a vertical cooler, where it is surrounded by cooling air that is provided by a secondary fan. The cooling air that is provided by a secondary fan reduces the temperature of that calcined dolomite that enters the cooler; the temperature of the calcined dolomite is cooled to below 100 °C, and the cooling air is correspondingly heated to over 700 °C so that it can be used as auxiliary combustion air for the combustion system. The cooled calcined dolomite is discharged by a vibrating unloader at the lower part of the vertical cooler, and it is transferred to a storage warehouse via a plate conveyor and bucket elevator.

The flue gas that comes from the pre-heater directly enters an electrostatic precipitator that is used to remove dust. The dust content of the exhaust gas after dust removal should be less than 80 mg/m^3 according to the relevant national requirements for environmental protection. The belt conveyor that is used for conveying dolomite is sealed with a belt corridor; the screening dolomite screening and the storage system used for the calcined dolomite are equipped with a pulsed bag filter. These measures effectively prevent environmental pollution that is caused by the storage and unloading of calcined dolomite and by the conveying and screening of dolomite.

(2)   Batching and pelleting: Calcined dolomite, ferrosilicon powder, and fluorite powder are weighed, batched, ground, and pelleted according to the process requirements. The specific process is as follows: The preparation of pellets is conducted to prepare raw materials for the reduction process. The calcined dolomite that is produced in a rotary kiln is sent to pre-crushing equipment via the bottom unloading device of the stock. The unloading device is something such as a rod valve and a vibrating feeder belt conveyor. The crushed calcined dolomite is then sent to the silo. The ferrosilicon powder is also crushed and sent to the silo, and fluorite is used directly in the batching process. The three materials are mixed in the batching machine according to the required proportions, fed via a belt conveyor into a ball mill for grinding, and then pressed into pellets.

The calcined dolomite is sorted first, and a jaw crusher is used to crush ferrosilicon from the raw material storage yard to a particle size of about 10–20 mm. The calcined dolomite, ferrosilicon, and fluorite are mixed in a white calcined dolomite:ferrosilicon:catalyst fluorite ratio of 100:7.8:0.06. The mixture is placed in a ball mill and milled into a mixed powder of about 120 mesh (120 μm). A bucket elevator is used to lift the ground powder to a pellet presser, and then, the powder is pressed under a pressure of 9.8–29.4 MPa into an spheroid pellet that is about 40 mm. The pellets are sieved, and then, the pellets and powder that is smaller than 30 mm are returned to the pellet presser again. The qualified pellets are conveyed to the reduction workshop.

(3)   Reduction: The pellets are heated to 1200 ± 10 °C in a closed reduction tank. To prevent the reduced Mg from being oxidized again under high temperature conditions, the reduction tank needs to be evacuated under a vacuum level of 13.3 Pa or higher. Magnesium oxide is reduced to magnesium vapor at this temperature and under this vacuum for 8–10 h; the magnesium oxide becomes crude magnesium after condensation. Specifically, the pellets are placed in a reduction tank of a reduction furnace and reduced by silicon in ferrosilicon to form metallic magnesium under vacuum. The high-temperature section of the reduction tank is made of heat-resistant steel, and the condensing section that extends out of the protective outer layer is welded with ordinary seamless steel pipes and placed horizontally in the reduction furnace, forming a structure that has "one furnace with multiple tanks".

The prepared pellets are placed in the reduction tank, which has a fire shield. The tank mouth is sealed, a jet pump is started to generate a vacuum, and the entire vacuum system is maintained below 5 Pa. The reduction furnace uses coal gas from the generating furnace as fuel. When the pellets are heated to 1150–1200 °C, the pellets are in their molten state. Under the catalysis of fluorite, silicon atoms in ferrosilicon reduce the magnesium ions of MgO to metallic magnesium. Metallic magnesium sublimates at high temperature into magnesium vapor. The vapor is cooled at the head region of the reduction tank and then condensed into solid crude magnesium. The general reaction time is 12 h. After the MgO in the pellet is completely reduced to metallic magnesium, the cover of the reduction tank is opened, and the metallic magnesium is removed using a hydraulic press. The remaining waste slag in the reduction tank is manually removed and poured into a fire pit; water is sprayed on the waste residue to prevent dust from rising. The main chemical reaction in silicothermic reduction is shown in (Scheme 1.3) [10]:

$$2MgO + 2CaO + 2Si(Fe) \xrightarrow{1190-1210\,°C} 2CaO \cdot SiO_2 + 2Mg \uparrow \qquad (1.3)$$

The reduction process is an intermittent operation. The specifications of the reduction tanks that are used by different magnesium producers are different. Accordingly, the amount of feeding material, magnesium output, and the reduction cycle are different. The general cycle is 8–12 h long. There are also large tanks that have a reduction cycle that is 24 h long. At the beginning of a reduction cycle, the pellets are first placed in the reduction tank, and this position corresponds to the region for the reduction reaction. A heat shield is then used to block the heat radiation of pellets, and a magnesium crystallizer and alkali metal trap are installed in turn. Finally, the end cover is fixed, and the system is evacuated.

To obtain dense crystalline magnesium, the reduction is carried out under a vacuum less than 10 Pa. The temperature in the region of the reduction reaction is maintained at 1473 K. At this temperature, the MgO · CaO in the pellets is reduced to metallic magnesium by silicon in ferrosilicon. The generated metal magnesium vapor escapes to the crystallizer. The vapor then crystallizes while the generated alkali metal vapor condenses in the alkali metal trap and then separates from the crystalline magnesium. After reduction, the vacuum unit is turned off, and the vacuum system is connected to the atmosphere. The reduction tank is opened, and the alkali metal trap, magnesium crystallizer, and heat shield are removed. The slag is cleaned and the crystallized magnesium is removed for the subsequent refining.

(4) Refining ingot casting: Crude magnesium is heated and melted. Then, it is refined via flux at a high temperature of about 710 °C, and the refined magnesium is cast into ingot, also known as refined magnesium.

(5) Acid-washing: The surface of magnesium ingot is cleaned with sulfuric acid or nitric acid to remove surface inclusions, and the surface is made to be more shiny.

2.1.4 Factors Affecting the Reduction of Magnesium

When magnesium is smelted via the Pidgeon process, vacuum thermal reduction is the core process and is also the process by which magnesium is obtained. The quality of this process directly affects the output of crude magnesium. Therefore, the factors that affect the reduction process must be addressed:

1. The effects of calcined dolomite's activity, ignition loss, and impurity content

The activity of calcined dolomite refers to the amount of active MgO in calcined dolomite and can be measured using volumetric measurements (YB/T105-2005). When the activity of calcined dolomite is between 30 and 35%, the magnesium output increases significantly. When the ignition loss of calcined dolomite is greater than 0.5%, it has a serious effect on the vacuum in the tank. At the same time, this also causes the formed H_2O and CO_2 to react with magnesium vapor and decreases the reduction rate. In addition, when the contents of other impurities, such as SiO_2 and Al_2O_3, are too high, they form slag with CaO and MgO, and this correspondingly reduces the activity of MgO. Meanwhile, impurities cause nodules in slag and then cause difficulties in operation. When the total amount of K_2O and Na_2O in the pellets is greater than 0.15%, oxidation-combustion loss occurs if there is metal magnesium in the reduction tank. This thereby reduces the actual yield of magnesium.

2. Effect of the reduction capability of ferrosilicon

Production practices have verified that the yield of magnesium is too low if ferrosilicon with a silicon content less than 50% is used. When ferrosilicon that has a silicon content above 75% is used in reduction, the magnesium output increases significantly. However, when the silicon content in ferrosilicon is further increased, the increase in magnesium output is not significant; thus, it is economical and reasonable to use ferrosilicon that has a silicon content above 75% to reduce calcined dolomite. Wang Yaowu et al. [11] invented an aluminum-containing reducing agent for use in magnesium smelting; they used aluminum ingots, waste aluminum, and aluminum alloys as raw materials to prepare the aluminum-containing reducing agent for magnesium smelting. The magnesium smelting process uses the same equipment as that used in the current Pidgeon process. Feng Naixiang et al. [12] invented a vacuum magnesium smelting method that uses a silicon-magnesium alloy as a reducing agent. The reduction reaction was carried out at 1000–13,000 °C, and a vacuum pressure less than 80 Pa was applied; this can greatly reduce the energy consumption of magnesium production and substantially increase the production efficiency. Han Fenglan et al. [13] invented a boron-containing mineralizer to replace fluorite. The amount of mineralizer used is less than that of fluorite, thereby reducing the environmental pollution of fluorine-containing compounds. Similarly, Han Fenglan et al. [14] used rare earth oxides as mineralizers to obtain modified magnesium slag that had better gelling activity. The application of such magnesium slag in cement and concrete blocks increases the addition ratio and enhances the utilization rate of magnesium slag. In terms of the reduction step, Dong Jiayu et al. [15] invented a new reduction step to modify the Pidgeon process. Materials were stirred via rotation of a reduction tank, so that the intermediate materials can be heated. This solves the problem of high vapor pressure in the materials' reaction. Thus, the reduction time is shorter, and the reduction temperature is lower.

3.   Effect of component ratio

Specific production practices have proven that with an increase in the molar ratio of Si/MgO, the yield of magnesium increases accordingly while the utilization rate of silicon decreases. To make a reasonable use of the reduction ability of ferrosilicon and to effectively increase the yield of magnesium, the Si/MgO ratio should be maintained in the range of 1.8–2.0. Meanwhile, the component ratio should be adjusted over time according to variations in the composition of the feed material and the content of ferrosilicon during the production process.

4.   Effects of reduction temperature and vacuum degree

In the normal production process, the reduction temperature should be controlled to be within the range of 1100–1150 °C, the furnace temperature should be controlled to be within the range of 1150–1200 °C, and the vacuum degree should be controlled to be between 10 and 15 Pa. If the reduction temperature is further increased, it has a great impact on the lifetime of the reduction tank and the furnace, although the reduction rate and recovery rate of metallic magnesium can be increased. Therefore, a further increase in reduction temperature should be considered in a more comprehensive manner.

5.   Effect of mineralizer

In the process of silicothermic reduction, according to the production conditions of dolomite and the reduction tank, 1–3% CaF_2 is added to the pellet materials for the purpose of accelerating the reaction between SiO_2 and CaO to form $CaSiO_3$, which then increases the reduction rate and the output.

6.   Effect of the particle size of the material

The particle size of calcined dolomite and ferrosilicon affect the formation of pellets as well as the reduction process of magnesium. Fine and uniformly-mixed pellets increase the contact area between calcined dolomite and ferrosilicon, and this subsequently accelerates the reaction and increases the reduction rate. However, when pellets are too fine, they are prone to thermal fracture and pulverization, and this affects the normal process of the reduction reaction.

7.   Effects of density and strength of pellets

Increasing the pelleting pressure can enhance the strength and density of the pellets. These characteristics can thus reduce crushing, increase the loading amount, enhance the thermal conductivity, accelerate the reaction, increase the output, and increase the actual yield of magnesium. The required density of the pellets is in the range of 1.9–2.1 g/cm^3; the strength of the pellets must meet the requirement that when a pellet falls freely from a height of 1 m to a cement floor, it breaks into 3–4 pieces without producing powder [10]. Wu Yong [16] pressed material into a honeycomb coal-like briquette to replace the walnut-like pellets that are used in the traditional process. This invention changed where briquettes are placed in filling the tank, increased the filling coefficient, and increased the single-tank filling. As a result, the heat energy distribution in the reduction tank is improved, the thermal efficiency is improved, and the output of a single tank is increased.

2.1.5 By-Products in Magnesium Production

1. Waste heat utilization of magnesium slag

After decades of development, especially with the application of regenerative reduction furnaces in the past ten years, the energy consumption of magnesium production has been reduced from 10 to 4–5 t with standard coal for 1 t of magnesium [17, 18]. However, if the energy consumption of producing ferrosilicon is included, the total energy consumption of the Pidgeon process is still as high as 8 t of standard coal. The unit energy consumption even exceeded that of metal aluminum production and was one of the nonferrous metallurgical industries that had the highest unit energy consumption [19]. The main reason for this is the low reduction rate of MgO during the reduction process, which was about 80% and has not been greatly improved upon to date [20, 21].

Recycle of waste heat generated via dolomite calcination: The dolomite calcination process consumes a lot of energy and produces high-temperature flue gas. If the flue gas is directly discharged into the air, the energy contained in the high-temperature flue gas is wasted. This energy waste can be reduced by recycling waste heat that is generated via dolomite calcination. Recycling waste heat can be implemented in the following ways: (1) High-temperature exhaust gas can be introduced into the vertical preheater to preheat the dolomite, and this can shorten the heating time of dolomite in the rotary kiln and reduce the amount of coal gas that is used. (2) High-temperature exhaust gas is used to generate electricity, and the generated electricity is used in various production projects or is connected to the grid. (3) A waste heat-recycling boiler can be added to collect the heat of flue gas and then to produce hot water for heating in winter or for use in daily life. The temperature of the final exhausted flue gas is greatly reduced, and it can be discharged into the air after dust is removed using cloth bags [22].

The temperature of the magnesium slag when it is just removed from the reduction tank is very high (at around 1200 °C), meaning that it contains a huge amount of energy. However, most magnesium enterprises have not taken measures to use this energy, and so the heat is wasted. Therefore, recycling waste heat of high-temperature magnesium slag effectively reduces the energy consumption in producing 1 t of magnesium. A recycling device for waste heat generated by magnesium slag can be developed to blow cold air upward from the bottom of the recycle chamber for magnesium slag, and high-temperature magnesium slag will simultaneously fall from the top of the chamber. The full heat exchange between magnesium slag and air produces high-temperature air, which can be used to preheat material or produce domestic hot water.

2.   Recycling and using magnesium slag

Magnesium smelting produces a large amount of magnesium slag. "For every 1 t of crude magnesium produced by an enterprise, 6–10 t of magnesium slag would be produced." [23] Magnesium slag has extensive applications, such as producing cement, concrete expansion agents, and paving materials. The Jingfu coal chemical enterprise in Fugu, Shaanxi uses magnesium slag and coke dust to produce nonburning bricks and obtained satisfactory results. This practice uses magnesium slag and also enables the enterprise to have a new source of profits [22].

2.2 Main Equipments in Magnesium Smelting via the Pidgeon Process

2.2.1 Preheating Furnace

The top of the preheater is generally equipped with a silo, and the silo is equipped with a level gauge to control the height of the material layer. There is a slip pipe between the silo and the preheater body to move dolomite into the preheater and to seal the material to prevent cold air from entering the preheater. After the raw materials are moved into the preheater, they flow under the action of high-temperature exhaust gas released by calcination and move in the direction opposite to the feed direction. The reverse flow carries out a sufficient heat exchange to uniformly preheat the materials. Various hydraulic push rods are used to push the preheated material into the kiln, and this can shorten the calcination time in the kiln. After the heat exchange, the flue gas can be processed by the dust collector and then discharged into the air.

2.2.2 Calcination Furnace

At present, the main equipment that is used for calcining dolomite includes the rotary kiln, shaft kiln, fluidized bed furnace, gas-fired vertical kiln, and composite vertical kiln.

(1)   Rotary kiln

The rotary kiln is the key equipment that is used in preparing active dolomite. The rotary kiln is mainly composed of a transmission device, supporting device with a roller, supporting device with a damper, barrel, kiln head, kiln tail, sealing device, kiln head cover, and combustion device. The barrel is a rotary part that is heated during the calcination of dolomite; it is made of a coiled and welded high-quality steel plate and is inclined to have a certain angle. The entire kiln body is supported by a roller, and there is a mechanical or hydraulic gear damper that can be used to control the axial displacement of the kiln body. The transmission device causes the barrel to rotate at the required speed using a gear ring that is in the middle of the barrel. In addition to the main drive, which is driven by a DC or variable-frequency speed-controllable main motor, the transmission part is also equipped with an auxiliary transmission device to ensure that the kiln body rotates slowly and prevents the kiln body from deforming when the main drive power is interrupted during installation and maintenance. To prevent cold air from entering the barrel and to prevent the overflowing of flue gas and dust from the barrel, a reliable kiln tail and kiln head composite with fish scale-like sealing devices are installed at the feed end (tail) and discharge end (head) of the barrel. In this design, a kiln with a larger diameter and shorter length is used; this reduces the vertical movement range of the kiln body, prolongs the ring-formation period inside the kiln, and saves space.

The particle size of dolomite ore is small (5–25 mm). When the kiln is rotated, the material inside the kiln rolls fully, and this intensifies the heat transfer via radiation. The materials are uniformly heated, and the calcination is complete. The calcination temperature is easy to control. When the rotary speed of kiln and the feed rate are reasonable, calcination produces calcined dolomite that has excellent activity and lower ignition loss. The raw material utilization rate of a rotary kiln (90%) is twice as large as that of a shaft kiln (40–50%). Considering the reduction in production costs, the rotary kiln is the most ideal choice of equipment for magnesium plants. Therefore, most domestic magnesium smelting manufacturers that use the Pidgeon process use a rotary kiln to calcine dolomite. At present, the largest domestic rotary kiln for calcining dolomite is Baosteel's φ 3 m × 70 m rotary kiln, which has a daily output of 600 tonnes of calcined dolomite. The smaller rotary kiln has dimensions of 1.2 m × 26 m and a daily output of 18 tonnes of calcined dolomite.

Previous production practices have proven that, the calcination effect of a rotary kiln is very remarkable as long as the process conditions are well controlled regardless of the structure of the dolomite. The calcined dolomite that is produced using a rotary kiln has higher activity, along with a higher extraction rate of magnesium and higher utilization rate of silicon. The calcined dolomite that is produced using other furnace types has high activity, and the extraction rate of magnesium and utilization rate of silicon are also relatively high. The activity of calcined dolomite produced using other furnace types is relatively poor.

(2) Shaft kiln

A shaft kiln is stationary vertical calcination equipment. At present, most domestic small-scale magnesium smelting plants that use the silicothermic method use this type of kiln; this particularly true of individual enterprises in townships and villages. Most of the magnesium plants that have an annual output of 200–300 tonnes use a shaft kiln as their calcination equipment. The shaft kiln is characterized by its simple structure and low one-time investment. Compared with the rotary kiln, the shaft kiln has lower output and larger loss and produces lower activity calcined dolomite. Because the cooling time of calcined dolomite is longer in a shaft kiln, the calcined dolomite easily absorbs water and is easily pulverized. There are other problems, such as a low temperature calcination zone, insufficient calcination, low thermal efficiency, large layer resistance, and a calcined material that has a wide distribution of particle size (60–150 mm). Therefore, the technology for calcining dolomite in the shaft kiln needed to be modified. In addition to improving the operation and selection of raw materials, other modifications include improving the structure of the furnace, increasing the mechanization level, using coal-fired machinery to reduce labor intensity, and using a semi-gas external combustion chamber with a bottom discharge. For current small magnesium plants, the rational use of shaft kilns is still essential.

(3)   Fluidized bed furnace

In recent years, fluidized calcination is a new technology that has been developed in China for calcining dolomite. The fluidized bed furnace has been used by foreign enterprises to calcine dolomite lime, and it is effective. Because of the low investment in equipment, large production capacity, and lower energy consumption than rotary kilns, domestic research has begun to adopt the fluidized calcination equipment and process. The basic process of fluidized calcination is as follows: Crushed small dolomite particles are added into the furnace, and combustion gas is introduced into the furnace. The materials are fluidized and stirred in the furnace to make the temperature of each material layer uniform and to efficiently decompose dolomite. During the calcination (which lasts about 15 min), the calcination temperature must ensure that there is no under firing or overfiring and that dolomite is completely decomposed. The decomposition temperature of $MgCO_3$ in dolomite is in the range of 734–835 °C and that of $CaCO_3$ is in the range of 904–1200 °C. CaF_2 is added to accelerate the decomposition of dolomite. The calcination time of dolomite is closely related to the calcination temperature and the particle size of dolomite. Fluidized bed calcination is a rapid calcination process at high temperature.

The calcination of dolomite in a fluidized bed furnace is ideal, and the investment is low. The calcination of dolomite in a fluidized bed furnace is new technology that is worth promoting, and it is especially suitable for medium-scale and small-scale magnesium plants.

2.2.3 Vertical Cooler

The high-temperature materials that are calcined in a rotary kiln flow into a vertical cooler that is encased by refractory materials. The cooler can be divided into four cooling-discharging regions, and the discharging speed within each region can be individually controlled according to the temperature of the materials. Center blast caps and cooling caps in each compartment are evenly distributed within the cooler. A blast cap is connected to a fan through a pipe. The material layer is stacked on and covers the blast cap, scattering down along the bus bar of the blast cap. The material makes reverse contact with cold air that is released through air holes in each layer of the blast cap, thereby completing the heat exchange. When the material is cooled to a temperature that is 80 °C higher than ambient temperature, the material is gradually discharged from the cooler under the action of a vibrating unloader. Heated air enters the rotary kiln directly from the kiln head cover and participates in combustion as secondary air. There are no moving parts in the cooler, and thus, it has a simple structure, good cooling effect, and less equipment maintenance.

2.2.4 Reduction Resort

In magnesium smelting via the Pidgeon process, the magnesium reduction tank is a very important component in the production process and is a consumable part. At present, the reduction tank is usually cast from high chromium-nickel alloy steel. The tank has a shorter service life (generally no more than 2–3 months), and the cost of the tank accounts for about 25% of the price per tonne of magnesium. The high cost of the reduction tank is a problem that magnesium smelters are concerned about and cannot do anything about. Therefore, producing a reduction tank that is higher quality and has a longer service life is of great significance for reducing the cost of magnesium production and for improving the economic benefits of enterprises [24] (Fig. 2.1).

The reduction furnace consumes huge energy. The traditional reduction furnace is the core equipment for silicothermic smelting of magnesium, and it is an externally-heated flame reverberatory furnace. A single row of reduction tanks is arranged horizontally in the reduction furnace. There are two dimensions of reduction tanks (φ 339 mm × 33 mm × 2000 mm and 370 mm × 35 mm× 2000 mm), and the amount of pellets in each tank is in the range of 165–180 kg. The reduction tank is made of thermally-resistant steel alloy. The reduction reaction of the pellets that are packed in the reduction tank occurs under vacuum (<5 Pa) and at a high temperature such as 1200 °C. Early reduction furnaces directly burned coal, and the furnace is manually fired. The consumption indicator of coal in the reduction is about 8 tce/t Mg; some furnaces have also used hot coal gas as fuel without preheated air, and the coal consumption in this reduction has been as high as 10 tce/t Mg or more.

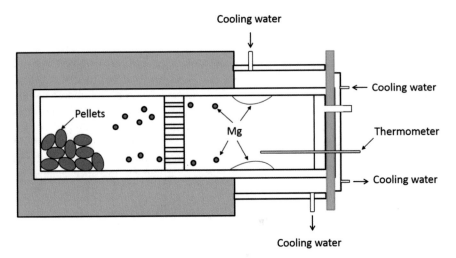

Fig. 2.1 Schematic diagram of reduction tank

From 1998 to 2004, the reduction furnace was changed from a single row of tanks to double rows of tanks. Meanwhile, a heat exchanger with metal dividing walls was used to preheat the air. The temperature used to preheat the air was about 450 °C, and the consumption indicator of coal in reduction was reduced to 6 tce/t Mg. At the same time, the waste heat from the flue gas after preheating the air is used to produce steam, and this can be used to drive the jet vacuum pump, thereby reducing the power cost of the vacuum. From 2004 to 2010, a regenerative magnesium reduction furnace with a horizontal tank was gradually developed and improved. The regenerative reduction furnace uses high-temperature technology to preheat air, and this can use clean energy such as low-calorific value fuels. Meanwhile, this furnace can be used to fully recover waste heat from the flue gas, greatly save energy, reduce consumption, reduce pollution emissions, and increase the output of the furnace and kiln [25, 26]. Therefore, heat storage technology has been rapidly promoted and applied in domestic reduction furnaces used for smelting magnesium. Also, the use of heat storage technology in the reduction furnace is becoming mature. The regenerative reduction furnace can preheat air and fuel to a temperature that is close to the working temperature of the furnace, and so it creates a uniform temperature in the furnace. As a result, the heating efficiency of the regenerative reduction furnace is much higher than that of traditional reduction furnaces and its coal consumption is reduced to 3.35 tce/t Mg; the coal consumption of a reduction furnace that has a more reasonable structure can be reduced to 3.0–3.1 tce/t Mg. However, compared to the theoretical energy consumption of 0.62 tce/t Mg in the reduction process, the energy consumption of the current reduction furnaces are still relatively high [27]. Table 2.1 compares the technical indicators and process parameters of a regenerative reduction furnace and a traditional reduction furnace.

Table 2.1 Technical indicators and process parameters of a regenerative reduction furnace and a traditional reduction furnace

Indicators and parameters	Traditional reduction furnace	Regenerative reduction furnace
Fuel consumption for producing a tonne of magnesium (t, using coal as fuel)	7	3
Temperature inside furnace (°C)	1200–1270	1200–1220
Material-magnesium ratio	6.8	6.2
Heating rate (C/min)	60	30
Cycle time (h)	11	10.5
Flue gas temperature (°C)	900	<150
Discharged volume of harmful gas (Bm^3/h)	2500	Trace amount
Preheated temperature of air and coal gas (°C)	N/A	1000

In 2011, Qu Tao et al. [28] used a semi-continuous feeding system to change the magnesium smelting reduction furnace to a semi-continuous vacuum-induction reduction furnace. The reduction region was heated via the vacuum induction method, and the reduction slag was discharged from the lower part of the reduction tank. This new design separated the reaction vessel from the device used to collect the magnesium vapor. This replaces the traditional discontinuous reduction process, increases output, and reduces energy consumption.

At present, most magnesium plants use horizontal reduction tanks. The horizontal reduction tanks cannot use the force of gravity on raw materials to drive automatic feeding, and so, manual feeding is generally used. In addition, the force of gravity on waste slag cannot be used to automatically remove slag, and so, manual slag removal is also required. Although some enterprises have achieved automatic slag removal from horizontal tanks, the failure rate of slag-removal equipment is relatively high, and magnesium slag is prone to remain in the reduction tank. After vertical tank technology was introduced, a reduction tank that is positioned vertically in the furnace makes feeding and slag discharge very convenient, and it easier to achieve automatic feeding and automatic slag discharge. These characteristics are conducive to the application of artificial intelligence and automation to production and can also reduce the cost of manual slag removal [22].

After 2010, some enterprises and scientific research institutes carried out a lot of research and development on energy savings and reduction of energy consumption in reduction furnaces. Technologies such as mechanized slag cleaners, vertical magnesium reduction furnaces, and composite magnesium reduction furnaces have been successively developed and tested in the magnesium industry. This new equipment has achieved good energy-savings and emission reduction effects.

1.   Mechanized slag removal and vertical reduction furnace for magnesium
     smelting

To reduce operating time of reduction furnaces, reduce ineffective heat loss, and
reduce labor intensity of workers, some magnesium smelting enterprises and scientific research institutes have developed mechanized slag cleaners for reduction
furnaces with a horizontal tank. This has mainly included spiral slag cleaners and
pneumatic slag cleaners. When a slag cleaner is used, operating a mechanized slag
cleaner is difficult and has a higher failure rate because of the low one-tank output of
the horizontal tank, more tank bodies being used, the tank body length being nearly
3 m (the high temperature section is 2 m long, and the vacuum section is nearly 1 m
long), deformation at high temperatures, material sticking to the tank and forming
a glaze, and other problems. Moreover, it is difficult to achieve mechanized feeding
with a horizontally-placed reduction tank, and so, mechanized operation cannot be
fully achieved. For this reason, the magnesium industry began to develop a reduction
furnace that uses a vertical reduction tank.
    Magnesium smelting technology that uses a vertical tank has a tank that is posi-
tioned upright in the furnace. Compared to a reduction furnace with a horizontal tank,
the vertical magnesium smelting reduction furnace has the following advantages [29]:
    Magnesium output of a single tank is increased. Raw materials in the vertical
reduction tank are filled fully and uniformly. The pellets are in contact with the inner
wall of the reduction tank, and this causes the pellets to be uniformly heated. Meanwhile, a more uniform temperature field is distributed in the tank, and this is favorable
for the reduction reaction. (2) The cycle of reduction production is shortened. Under
the action of gravity, the transfer of pellets and crude magnesium and automatic
discharge of slag are achieved; this shortens the auxiliary operation time of reduction production and reduces ineffective heat loss caused by the reduction furnace.
(3) Utilization of waste heat that is generated by reduction slag is strengthened. Slag
in the vertical reduction tank is automatically discharged to the slag box under the
action of gravity. The waste heat of high-temperature slag is recycled through a heat
exchange device, and the recycled heat is used to preheat the reduction material and
to achieve the use of the waste heat of the slag material. (4) The service life of the
reduction tank is extended. The reduction tank stands upright in the heating furnace
body, and its placement direction is consistent with the direction of gravity. Thus, the
force on the tank is uniform, and the reduction tank does not easily deform; a rapid
change in the surface temperature is alleviated through mechanized feeding, slagging,
and preheating of material, and this extends the service life of the reduction tank.
(5) Working conditions are improved. Mechanization of taking magnesium, adding
materials, and discharging slag improves working conditions and reduces labor intensity. After several years of technology development and production practice, it has
been shown that the reduction energy consumption of the vertical tank reduction
furnace can be reduced to 2.4–2.5 tce/t Mg. Although erecting the reduction tank
does not fundamentally solve the problems of the low heat transfer efficiency of the
silicothermic method and the "chimney effect" of the vertical tank exacerbates the
working environment. However, compared to the horizontal tank reduction furnace,

the vertical tank reduction furnace is more conducive for large-scale and mechanized development and has greater research and promotion application value.

Among them, a new magnesium smelting technology that uses a composite vertical tank has been developed by a magnesium-based technology and materials research and development team that is jointly associated with Zhengzhou University, Xi'an Jiaotong University, and other institutes. Through more systematic thermo-chemical research, they correct, supplement, and improve the relevant basic theories of the silicothermic method used in magnesium smelting. Their work has Aimed at the problems of using a horizontal tank or vertical tank in the silicothermic method, and they have developed an innovative new type of magnesium smelting process that uses a composite reaction furnace as the core equipment. In addition, a steam generator that uses crystallization heat and a recycling device used for waste heat of reduction slag are key equipment. The new method is a refining-free "two-step" magnesium smelting process that uses a composite vertical tank [29]. The large-scale composite vertical tank consists of three independent parts: a crystallizer, reducers, and a slag cleaner. The longitudinal and radial directions are composed of multiple independent components with perfect structure and complete functions; the combination forms a composite structure. The magnesium output of each composite vertical tank can reach 800 kg/d, and the service life can reach about 300 d. There are 50 composite vertical tanks in a reduction furnace. Feeding, slagging, and magnesium discharging are all mechanized. The flue gas and fuel system are self-balanced, and their operations are automatically controlled. Thus, the reduction furnace can be used for large-scale and automated production. This technology has been built and put into production with a output of 12,500 t/a. In actual operation, the comprehensive energy consumption of an entire plant that uses this technology is about 3 tce/t Mg, and the reduction time is about 6 h, which is only 2/3 of that used in the Pidgeon process. A single crystallizer can produce approximately 200 kg of magnesium at a time, and the purity of the crystallized magnesium is higher than 99.8%. These observations indicate that the magnesium smelting process has achieved significant technological progress [27].

At present, the most current magnesium reduction technology in the world is the new technology of vacuum smelting of metallic magnesium using internal electrothermal heating. This technology uses electricity as a secondary energy source to heat reaction material from the inside. This technique avoids pollution problems caused by the combustion of materials, increases the reduction temperature, accelerates the reaction process, shortens the reduction cycle, enhances thermal efficiency, and reduces energy consumption. Additionally, it increases the furnace capacity and increases the output of a single furnace. This process does not require expensive metal reduction tanks, and the cost is reduced. Also, it can be designed as continuous production with automatic feeding and automatic discharging, which greatly reduce labor intensity [1].

2.2.5 Heating System of a Reduction Resort

Most magnesium plants in China use coal as fuel. Every 1 t of magnesium produced consumes 8–12 t of high-quality coal. This wastes energy and resources, and also emits a large amount of waste gas, causing serious environmental pollution. From a comparison of the use of clean energy by advanced magnesium smelting enterprises at home and abroad with common magnesium smelting enterprises in China, it is found that the use of clean energy is an effective way to improve magnesium smelting via the Pidgeon process.

The thermal reduction reaction of magnesium smelting can be replaced by microwave heating. Specifically, microwave heating does not use heat conduction to heat but uses high-frequency electromagnetic radiation to act on polar molecules in a material. The violent movement and collision of polar molecules generates heat that can heat materials. In addition to uniform heating, microwave heating is convenient to use to start, control, and stop the heating process, and the temperature can be controlled through adjusting the frequency. Microwave heating can minimize heat loss because the reduction tank can be made of materials that reflect microwaves, thus minimizing microwave absorption. As reported in the literature, "Compared with traditional heating methods, microwave heating has significant advantages such as high efficiency, uniform heating, clean and pollution-free, rapid start and stop heating, easy to control, and improving material properties" [17]. In promoting the application of microwave heating in the thermal reduction of magnesium, the main difficulties are as follows: the conversion efficiency from power to microwaves needs to be improved, and power-microwave conversion equipment with large power still needs to be developed. Although microwave heating has not been applied in the field of magnesium smelting, there have been breakthroughs in the application of microwave heating technology in other metallurgical technologies [18].

2.2.6 Auxiliary Devices

1. Automated equipment used in magnesium smelting

Improvements to the levels of automated monitoring and process control of magnesium smelting can be used to optimize the process flow, enhance the efficiency of magnesium smelting, and decrease entrepreneur's expenses. Moreover, these improvements can effectively reduce the time that workers are exposed to harmful working environments. In addition, the monitoring system that we studied adds units for utilizing waste heat and treating waste gas; thus, effective monitoring of waste heat utilization can be achieved, and it can be ensured that discharged exhaust gas always meets environmental protection requirements.

(1)   There are three main approaches for enhancing the mechanization and automation levels of the Pidgeon process. One is the use of a closed joint line that

consists of microcomputer batching, grinding, and pelleting, and the second is the use of mechanized slag removal, tank cleaning, feeding, and magnesium extraction equipment. The closed joint line uses mechanized continuous casting. Advanced foreign enterprises have achieved all three of these points, whereas the current advanced domestic enterprises have only achieved two aspects. Most magnesium plants only achieve one or even none at all, and this indicates that the mechanization level of the enterprises is very low and that improvement is urgently needed.

Siemens WINCC configuration software is used to achieve real-time automatic monitoring of the entire magnesium smelting line. The Siemens S7-1200 or 1500 series PLC is used to carry out temperature and logic controls on the entire magnesium smelting system. When designing the monitoring system, the following aspects are the main concerns:

(1) Utilization of waste heat generated via the calcination process. High-temperature exhaust gas is used to preheat raw materials or to produce hot water. A temperature sensor is used to measure the temperature of the exhaust gas, the temperature of the exhaust gas after waste heat utilization, and the temperature of the raw materials after preheating. The temperature data are sent to the PLC for centralized real-time monitoring. An appropriate sensor should be selected according to the range of the measured temperatures. In principle, the error should be as small as possible, and the price is relatively low. If high-temperature exhaust gas is used to produce hot water, hot water boilers need to be added. Siemens WINCC + PLC can be used to design a set of separated monitoring subsystems for boilers, and these provide more convenient control.

(2) Treatment of exhaust gas. When purifying and removing dust from the exhaust gas after the waste heat is used, it is necessary to determine the temperature of the gas. For gas at a higher temperature, the temperature must be lowered below the maximum temperature that the filter material can withstand. Generally, it should be controlled to be below 120 °C, and cloth bags can be used to remove dust. The cooling methods that are used during bag dust removal include forced air cooling, water cooling, and natural air cooling. The control scheme is determined according to different cooling methods. In this stage, concentration sensors are necessary for monitoring various harmful gases. If the content of emitted harmful gas is within the scope of environmental protection requirements, it is normally discharged. If it exceeds the standard, a warning message is issued to remind the production personnel to take measures to intervene until the requirements are met.

(3) Effective monitoring of furnace temperature, furnace pressure, and flue gas temperature of the reduction furnace. The furnace temperature and temperature of flue gas can be measured using a nickel-chromium and nickel-silicon thermocouple, and the vacuum inside the furnace is measured to determine the furnace pressure. The measured data are input to the PLC host module

through the PLC analog input module. The CPU of the PLC host receives the opening magnitude signal of the combustion valve and the speed control signal of the vacuum pump through PID calculation; the flame size is subsequently adjusted, and the required vacuum degree in the furnace is maintained [22].

2.   Automated slagging technology

At present, most of the magnesium plants in Fugo, Shaanxi still use manual slagging. Enduring a long-term harsh environment, which includes dust and high temperature, has seriously harmed the health of workers. There are fewer and fewer people engaged in slagging work although wages continue to increase. Some magnesium plants use forklifts to dig out the slag, and this enhances slagging efficiency and reduces the physical expenditure of workers. However, forklift drivers are still in a dusty and high-temperature environment that is not good for their health. Tan Yuhao et al. [30] invented dust-free automatic slagging equipment for use with magnesium smelting via the Pidgeon process. The machine can automatically remove dust and slag, and this enhances production efficiency. Furthermore, the machine has a slag-collecting device that is convenient for the secondary use of slag. Liang Xiaoping et al. [31] invented a feed device for magnesium smelting that is conducted via the Pidgeon process. The device is equipped with a trolley that moves longitudinally on a horizontally moving trolley. A spiral or air flow slag cleaner is installed on the longitudinally moving trolley, and the longitudinally moving trolley is raised and lowered relative to the horizontally moving trolley. This device has a higher feeding efficiency and effectively reduces the production costs and labor intensity of the workers in the magnesium smelting industry.

References

1. Junping D, Jingxin S, Xiaogang W, Jianxun R, Xinwei T, Zimin F, Junfeng Q (2007) Study on temperature rule of inner heat magnesium smelting vacuum furnace. In: Proceedings of 2007 high tech new material industry development seminar and annual meeting of editorial board of material guide
2. Xiaosi S (2011) Overview of magnesium production process. Shanxi Metall 03:1–4
3. Juan L, Wang J, Bi Z et al (2009) Evaluation method for clean production of magnesium industry with Pidgeon process. Environ Sci Technol 32(8):176–178
4. Jinping L, Xuechun Y, Shuishui X (2005) The deficiency and improvement methods of Pidgeon process Mg-smelting. Metall Energ 24(5):21–23
5. Rong JG (1999) Review on calcination of dolomite by Pidgeon process. Light Metals 10:36–38
6. Kai Y, Junxiao F, Zhibin S (2006) Review of industrial kiln development of Dolomite calcination for Pidgeon' s magnesium reduction. Ind Furnace (06):16–19
7. Kun L (2002) Process characteristics and arrangement of Beckenbach annular shaft kiln. Wuhan Iron Steel Corporation Technol 01:51–55
8. Qibo C, Junxiao F, Zhibin S (2006) Numerical simulation on flow field of the new type regenerative magnesium reducing furnace. Ind Heating (06)
9. Juan L (2008) Study on establishment of the cleaner production index system and its evaluation methods of mangesium industry with Pidgeon process. Jilin University, Changchun

10. Xutao W (2010) Experimental investigation on reaction characteristics for magnesium residues as sorbent. Taiyuan University of Technology, Taiyuan
11. Yaowu W, Jianping P, Yuezhong D (2015) Preparation method and use method of aluminum-containing magnesium-smelting reducing agent. China, Patent No: cn 104789775, Publication date: 2015.07
12. Naixiang F, Yaowu W, Jianping P, Yuezhong D, Daxue F (2013) Vacuum magnesium making method using magnesium-silicon alloy as reducing agent. CN 102864315 a, Publication date: 9 January 2013
13. Fenglan H, Laner W, Qixing Y, Chun D, Shengwei G, Youjun L, Yuhong C (2012) Pidgeon process for making magnesium and boronic mineralizer for partially replacing fluorite. CN 102776387 a, Publication date: 14 November 2012
14. Fenglan H, Laner W, Qixing Y, Chun D, Shengwei G, Youjun L (2012) Pidgeon magnesium smelting process and applications by taking rare earth oxide (REO) as mineralizer. CN 102776388 a, Publication date: 14 November 2012
15. Jiayu D, Xiurong W, Jing W, Wanlan T (2012) Novel reduction method in Pidgeon magnesium refining process. CN 102409185 a, Published on: 11 April 2012
16. Yong W (2014) Improved Pidgeon-process magnesium refining process. CN 103602833 a, Publication date: 26 February 2014
17. Minić D, Manasijević D, Đokić J et al (2008) Silicothermic reduction process in magnesium production. J Therm Anal Calorim 93(2):411–415
18. Liangliang S (2009) Benefit comparison and analysis on magnesium smelting by electrolytic method with Pidgeon method. Energ Saving Nonferrous Metal 25(5):6–15
19. Cherubini F, Raugei M, Ulgiati S (2008) LCA of magnesium production technological overview and worldwide estimation of environmental burdens. Resour Conserv Recycl 52:1093–1100
20. Wang YW, You J, Peng JP et al (2016) Production of magnesium by vacuum aluminothermic reduction with magnesium aluminate spinel as a by-product. J Miner Metals Mater Soc 68(6):1728–1732
21. Tieyong Z (2008) Potential and countermeasures for developing circular economy in China's original magnesium industry. Renew Resour Circ Econ 1(9):1–6
22. Rui S (2018) Process status and energy saving optimization analysis of Pijiang magnesium smelting process. China High Tech Zone 10:162–163
23. Qijun Z, Yuxin L, Yufeng W (2011) Status of recovery and utilization of magnesium slag. Renew Resour Circu Econ (06)
24. Junkang J, Guangdong Z (2017) A vertical condensation crystallization ware for producing magnesium metal. China, Patent No. cn 206538461, Publication date: 2017, 10, 03
25. Dongmei L, Ruitang C, Guimin C et al (2008) Discussion on energy saving technology of accumulation heated magnesium reducing furnace. Energ Saving Nonferrous Metall 2:37–40
26. Dehong X (2007) Several problems in the application of regenerative combustion technology in magnesium smelting process. In: Proceedings of national magnesium industry conference, Nanjing, pp 41–46
27. Dongmei L (2018) Current status and development of reduction furnace energy saving technology in Pidgeon smelting magnesium. Energ Saving Nonferrous Metall 10(5):15–19
28. Tao Q, Bin Y, Yang T, Yongnian D, Hongxiang L, Dachun L, Baoqiang X, Buzheng Y, Wei L, Wenhui M, Qingchun Y, Yifu L, Heng X, Yong D, Wenlong J, Xiumin C, Yaochun Y, Qi L (2011) Semi-continuous vacuum induction heating magnesium reduction furnace. CN 201942729 u, Publication date: 2011.08.24
29. Jilin H, Shaojun Z (2014) Energy saving and emission reduction and development potential of magnesium smelting. In: Proceedings of the first academic conference on nonferrous metal metallurgy in China, Beijing
30. Yuhao T, Zhanyi C, Liping L (2016) Automatic dust-free slag removing machine for smelting magnesium through Pidgeon process. China, Patent No. cn 105571330 a, Publication date: January 11, 2016
31. Xiaoping L, Xiaojun L (2014) Loading device for smelting magnesium by Pidgeon process. CN 104195353 a, Publication date: December 10, 2014

Chapter 3
Magnesium Slag Generated by Reduction Smelting Using Pidgeon Process

Abstract The generation process and behavior of magnesium reduction slag via Pidgeon process are described in this chapter. Main compositions, phases, particle size distribution and harmful elements in magnesium slag were analyzed. The state and development of the techniques of treatment and effective use of magnesium slag were reviewed. The advantages and technical challenges of integrated utilization of magnesium slag were proposed.

Keywords Magnesium slag · Pidgeon process · Dicalcium silicate · Fluorine pollution · Dust pollution

3.1 Composition and Phases of Magnesium Slag from Reduction Smelting

3.1.1 Generation of Magnesium Slag

During magnesium metallurgy via the Pidgeon process, pellets of raw materials react in a reduction retort at high temperature and under negative pressure. Magnesium in dolomite is reduced, becomes magnesium vapor. Under vacuum, magnesium vapor reaches a water circulation-cooling section of the reduction retort and condenses into crude magnesium ingots. After the reaction is complete, the discharge end of the reduction retort is opened, and the magnesium ingot and potassium/sodium crystallizer are removed. Slag is manually or semi-mechanically discharged. The slag remains pellet shape at high temperature, and is naturally cooled and pulverized after leaving the retort. The final slag is a powder that has a gray-white appearance, and mixed with granules that are not completely pulverized. The slag is manually transported to a slag yard for further cooling. Sometimes, water is poured on the slag to accelerate cooling. However, water spray aggravates dust flying off of the magnesium slag. For magnesium metallurgy via the Pidgeon process, the discharge of the slag and an example of the current situation of a slag field in Hongguozi production area of Huiye Magnesium Co. Ltd., Ningxia is shown in Fig. 3.1.

Magnesium slag that is piled in the open air is basically in the form of fine powder after the magnesium slag decomposes naturally. The particle size of the

© The Author(s) 2021

L. Wu et al., *Comprehensive Utilization of Magnesium Slag by Pidgeon Process*, SpringerBriefs in Materials, https://doi.org/10.1007/978-981-16-2171-0_3

(a) Slag tapping   (b) Slag yard

Fig. 3.1 Huiye Magnesium Co. Ltd., Ningxia (2009)

slag is measured with a laser particle size tester (Microtrac X-100), and the particle size distribution diagram is shown in Fig. 3.2. Table 3.2 shows the corresponding composition of the slag. As seen in Fig. 3.2, the median diameter of magnesium slag that is placed for a long time is 15–16 µm, and the particles with a size is in the range of 1–10 µm account for about 40% of the volume fraction. These tiny particles are the ones that cause harmful dust.

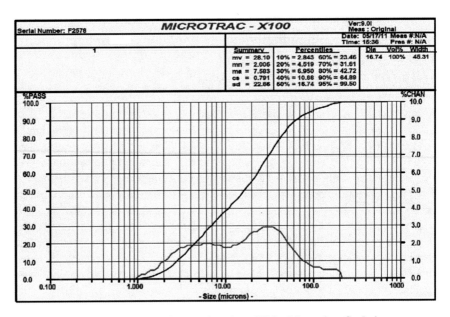

Fig. 3.2 Particle size distribution of magnesium slag of Huiye Magnesium Co. Ltd

3.1.2 Composition and Phases of Magnesium Slag

The chemical components of magnesium slag include calcium silicate, silica, calcium oxide, and magnesium oxide and so on. From random sampling in the slag yard of Huiye Magnesium Co. Ltd., the detected compositions of magnesium slag are shown in Tables 3.1 and 3.2.

The composition of magnesium slag of Sun Magnesium Enterprise is shown in Tables 3.3.

The composition data of magnesium slag are different in different production batches. They are also different when different sources of raw materials are used. MgO content reflects if the reduction reaction is complete, and it is directly related to the product yield and heating time during reaction. Heating time is closely related to energy consumption and production costs, and so, sufficient attention must be paid to heating time. It is generally considered that MgO content should be controlled below 5%. The data of MgO in the magnesium slag of Sun Magnesium Enterprise is obviously high, and the specific reason for this is unknown.

The phase of magnesium slag is relatively complicated, and it is related to the details of the raw materials and the production process. However, the main phase is dicalcium silicate (Ca_2SiO_4, abbreviated as C_2S). The X-ray diffraction patterns of the magnesium slag generated by Ningxia Huiye Magnesium Co. Ltd. and Ningxia Sun Magnesium Enterprise are shown in Figs. 3.3 and 3.4, respectively.

Table 3.1 Composition of magnesium slag of Hongguozi Section of Huiye Magnesium Co. Ltd (%)

Composition	Slag 1	Slag 2	Slag 3	Slag 4	Mean
SiO_2	29.37	32.6	31.33	38.6	32.98
CaO	51.73	43.98	45.03	54.85	48.90
MgO	6.41	4.62	4.67	3.21	4.73
CaO/SiO_2	1.76	1.35	1.44	1.42	1.48

Table 3.2 Composition of magnesium slag generated by Fugu Section of Huiye MagnesiumCo. Ltd. (%)

Composition	MgO	CaO	SiO_2	P_2O_5	Fe	Al	Mn	Na
%	5.12	42.35	26.78	0.061	3.85	0.604	0.061	0.979
CaO/SiO_2	1.58							

Table 3.3 Composition of magnesium slag of Sun Magnesium Enterprise, Ningxia

Composition	CaO	SiO_2	Al_2O_3	Fe_2O_3	MgO	K_2O	Na_2O	SO_3	TiO_2
%	46.13	26.54	1.53	5.94	11.82	0.11	0.09	0.48	0.01
CaO/SiO_2	1.74								

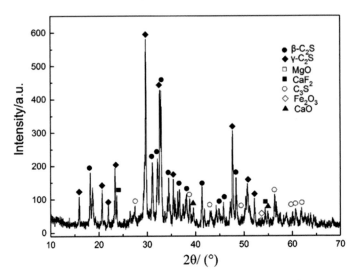

Fig. 3.3 XRD pattern of magnesium slag generated by Huiye Magnesium Co. Ltd

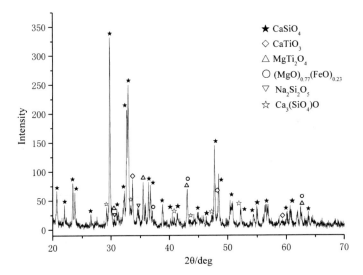

Fig. 3.4 XRD pattern of magnesium slag generated by Sun Magnesium Enterprise

3.2   Harmful Components in Magnesium Slag

3.2.1   Fluorine Pollution

Magnesium metallurgy via the Pidgeon process requires using calcium fluoride (CaF_2) as a mineralizer, and the amount of CaF_2 that is commonly added is about 2–3%. Calcium fluoride does not participate in the reduction reaction; it only promotes the reaction. After the reduction reaction is complete, the calcium in the CaF_2 becomes a component in the slag, and the fluorine becomes gas at high temperature and forms HF in the presence of water vapor. HF is discharged to the atmosphere through a vacuum exhaust line. Some of the calcium fluoride condenses into solid at the potassium/sodium crystallizer, in the water-cooling section of a reduction retort and enters into slag. Based on the data provided by the China Magnesium Industry Association, the pollutant emissions that are generated via magnesium metallurgy using the Pidgeon process are estimated [1]. In addition to the coal-combusting pollution, the main pollutants that are generated during the process include CO_2, SO_2, HF, and particulate matter (dust). According to the literature, every tonne of magnesium that produced via the Pidgeon process produces 24–34 kg of HF and 85–110 kg of particulate matter [1]. A scaled-down pilot Pidgeon process was carried out to reduce magnesium with double reduction retorts that each held 10 kg of pellets. The migration of fluorine in the experiment was studied, and the thermodynamic software FACTSAGE was used to verify the results [2]. After excluding the experimental error and taking the average value, it was found that more than 90% of the fluorine that was in the raw materials ultimately entered the magnesium slag [2, 3].

China is the world's largest magnesium producer. In 2007, China's primary magnesium output accounted for more than 80% of the world's total output [4]. A comparison of China's magnesium output to the world's magnesium output from 2001 to 2007 is shown in Fig. 3.5 [4].

At present, magnesium metallurgy via the Pidgeon process accounts for nearly 95% of the total magnesium production in China [5]. Except for a small amount of magnesium slag that is recycled, most of magnesium slag is discarded as waste. With the accumulation or landfill of a large amount of magnesium slag, hazardous substances that are in magnesium slag dissolve out in the rain. The flowing of hazardous substances into rivers and lakes has a serious impact on crops and the surrounding environment, and this seriously endangers human health and crop growth. The *"European Landfill Directive 1999/31/EC"* sets fixed emission limits for the emission concentration of various pollutants. If the concentration of a pollutant in waste exceeds the limit, direct landfill of the waste is prohibited, and the waste must be pretreated. However, if the concentration of the pollutant that is discharged in the waste is lower than the limit, it is considered that the waste will not cause great harm to the environment and that it can be directly landfilled without pretreatment. The emission limits of some important elements in this standard are shown in Table 3.4.

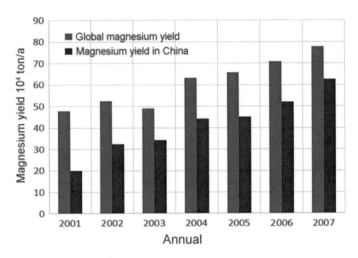

Fig. 3.5 Magnesium output (10^4 tonne) of China and worldwide

Table 3.4 Emission limits of some important elements in the "*European Union Waste Landfill Guidelines*"

Elements	Cr	Mo	Cu	Cd	F	Pb	Zn	Ni
Limit (mg/kg)	0.5	0.5	2.0	0.04	10.0	0.5	4.0	0.4

Professor Han Fenglan of North Minzu University conducted leaching experiments on magnesium slag samples using to the standard procedure in EU EN 12457-2 "Characterization of Waste-Leaching; Compliance test for leaching of granular waste materials and sludges", and the results are shown in Table 3.5 [6].

As seen from the test results shown in Table 3.5, the fluorine in magnesium slag that is generated via the Pidgeon process seriously exceeds the standard limit. Thus, the slag cannot be landfilled directly because it will cause serious consequences.

Table 3.5 Leaching experiment results of magnesium slag generated by Huiye Magnesium Enterprise (measured according to EU standard)

Element	Ca	Mg	Na	K	F	S	Al	pH	Conductivity (μs/cm)
Test value (mg/kg)	1970	0.9	18.7	16.8	**72.9**	30.4	84.3	13.4	1764
Element	Cr	Mo	Cu	Cd	Ni	Pb	Zn		
Test value (mg/kg)	0.042	0.207	0.017	0.0005	0.005	0.002	0.047		

Table 3.6 Leaching results of magnesium slag sample based on China's national standard

Initial leaching solution	Distilled water with pH = 6.9	Distilled water with pH = 3.2 (pH adjusted using sulfuric acid and nitric acid)
Fluorine content in leaching solution (mg/kg)	139.0	139.5

Similarly, with reference to HJ/T299-2007 "Solid waste-Extraction procedure for Leaching Toxicity-Sulfuric acid and nitric acid method" issued by the Ministry of Environmental Protection of China, the magnesium slag was leached, and the results are shown in Table 3.6.

The maximum concentration of fluorine emission as defined in the national standard GB5749-2006 for drinking water safety is 10 mg/kg. The detected fluorine content in the leaching experiment of magnesium slag was as high as 14 times of the limit. On the basis of the above experimental results, the following conclusions can be drawn: the fluorine content in magnesium slag exceeds the maximum limit several times according to the EU standard, or more than 10 times greater than the maximum limit, according to the China's national standard. Therefore, direct stacking and landfilling of untreated magnesium slag can easily cause fluorine leaching from the magnesium slag, and this poses a great threat to drinking water.

3.2.2 Dust Pollution

As seen from Figs. 3.1 and 3.2, after the reaction is complete and the slag is discharged at high temperature, the magnesium slag that is naturally cooled and weathered contains a large number of tiny particles that have a size range of 1–10 μm; this causes dust to spread in the storage location. The spread of dust pollutes the environment and makes collection and transportation inconvenient. The harms that dust has on the human body include: (1) Hard and sharp-shaped dust can cause mechanical damage to the human respiratory mucosa. (2) Long-term inhalation of a certain amount of dust can result in dust gradually depositing in the lungs, causing the lungs to produce progressive and diffuse fibrous tissues; the resultant respiratory disease is called pneumoconiosis. (3) Inhalation of a certain amount of silica dust hardens lung tissue and causes silicosis. In addition, toxic dust that mainly containing sulfur dioxide flue gas has become the main factor that affects the air environment in China. China's national standard classifies the dust that is generated at an industrial production site into total dust and respirable dust. Total dust is defined as "dust that can enter the entire respiratory tract (nasopharynx and throat, thoracic bronchus, bronchioles, and alveoli)", and respirable dust refers to dust that is deposited in the alveolar area. The British Medical Research Council (BMRC) proposed in 1952 that the deposition efficiency of particles with an aerodynamic diameter of 5 μm is 50%, and the deposition efficiency of dust with an aerodynamic diameter greater than 7.07 μm is 0. The American Conference of Governmental Industrial Hygienists (ACGIH) stipulates that the deposition ratio of dust particles with an aerodynamic diameter of 3.5 μm is 50%, and the deposition ratio of dust particles with an aerodynamic diameter greater than 10 μm is 0. BMRC regulation of respirable dust is used in China. Dust particles that are larger than 10 μm usually fall under the action of their own weight, whereas dust particles that are smaller than 10 μm easily form flying dust, smoke, and smog. These particles remain in the air for a long time and are easily inhaled by the human body, causing severe harm. Two of the main components in magnesium slag are SiO_2 and CaO. Free silica dust has been listed as a human carcinogen by the International Cancer Research Center. Dust that contains these substances may cause tumors in the respiratory system or other tissues.

3.3 Current Status of Comprehensive Utilization of Magnesium Slag

Statistics show that the output of China's magnesium industry was 873,900 tons in 2014 and 863,900 tons in 2018. The outputs of magnesium in China from 2014 to the first half of 2019 are shown in Fig. 3.6 [7].

The Pidgeon process produces 7–8 tonnes of reduced slag for every onetonne of magnesium that is produced. The rapid development of China's magnesium industry is accompanied by the emission of a large amount of magnesium slag. A large

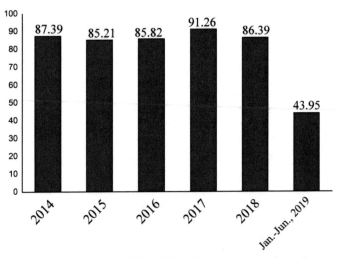

Fig. 3.6 Outputs of China's magnesium industry from 2014 to the first half of 2019. *Data source* China nonferrous metals industry association

amount of magnesium slag is still piled up or landfilled. This occupies, a lot of precious land, and also has a great impact on the surrounding environment; it can seriously endanger human health and the growth of animals and plants. Therefore, how to fully utilize magnesium slag has become a limiting factor in the sustainable development of the magnesium industry in China. At present, research hotspots regarding the recycling and reuse of magnesium slag in China mainly focus on the following aspects: using magnesium slag as a desulfurizer, replacing lime as a slag-making agent in steelmaking, preparing cement clinker and using it as an additive in mixtures with other materials to activate cement, preparing building materials, preparing fertilizer and preparing geopolymers.

3.3.1 Magnesium Slag as a Desulfurizer

With the rapid development of economic construction, China's coal-dominated energy consumption increases productivity and creates huge material wealth, but it has also brought a large amount of sulfur dioxide emissions and caused a tremendous strain on the environment. At present, the most frequently used desulfurizers in industry are limestone and dolomite. The principle is to use alkaline oxide or carbonate to react with sulfur dioxide and then to generate sulfate or sulfite to achieve the purpose of sulfur fixation. The amount of CaO in magnesium slag is as high as 40–50%, which is a favorable condition for using magnesium slag as a desulfurizer. The research team of Fan Baoguo of Taiyuan University of Technology carried out a large number of experimental studies on the desulfurization performance of magnesium slag[8–15]. Zhao Jianli et al. [16] and Xiao Yongqiang et al. [17] also reported relevant studies. Xu Xiangbin et al. used magnesium slag as a sulfur fixing agent for fire coal and an industrialization test was carried out for the desulfurization of molten iron that is produced by iron and steel plants. When magnesium slag was used as desulfurization agent in the steel industry, a desulfurization efficiency over 80% was achieved [18, 19].

3.3.2 Preparation of Cement and Construction Materials via Magnesium Slag

The team led by Cui Zizhi of Ningxia University carried out a series of studies on the preparation of magnesium slag concrete via the mixing of magnesium slag and other solid wastes and studied the strength, carbonization characteristics and shrinkage mechanism of the prepared magnesium slag concrete [20–22]. Chen Guanjun [23], Ji Ying et al. [24] of Xi'an University of Architecture and Technology studied the properties of controllable expansive cementitious materials and cement cementitious materials that were prepared using magnesium slag. Ji Guangxiang of

Chongqing University studied the preparation and performance of alkali-magnesium slag concrete without the use of an autoclave and aeration [25]; Peng Xiaoqin et al. studied the preparation of magnesium slag bricks [26]. Luo Feng and Xiao Liguang of Jilin Institute of Architecture and Engineering studied the preparation mechanism of cementitious materials that contain magnesium slag and the preparation of magnesium slag wall materials [27–30]. Fang Renyu et al. tested the use of magnesium slag instead of limestone to calcinate cement clinker. The results showed that at a calcination temperature of 1450 °C and with a calcination time of 30 min, the properties of the clinker met the standard when 15% of the limestone content was replaced by magnesium slag. Mechanism analysis indicated that this result is related to the facts that there are higher contents of CaO and SiO_2 in magnesium slag and that the main mineral component is C_2S [31]. Professor Han Fenglan of North Minzu University used magnesium slag and manganese electrolytic slag to prepare sulphate aluminum cement clinker. Manganese electrolytic slag was found to contained dihydrate gypsum, and magnesium slag contained C_2S. Moreover, the main chemical components of magnesium slag are Fe_2O_3, SiO_2, Al_2O_3, and CaO, and this is consistent with the main chemical components in sulfate cement clinker. Calculations determined that the content of manganese electrolytic slag and magnesium slag in the raw material can each be as high as 21%. The best sintering temperature for the raw materials is 1260 °C, and the holding time is 30 min. The mineral phase of the sample that was sintered at these conditions mainly contains C_2S and C_4A_3S. The addition a certain amount of gypsum to the prepared cement clinker enhances the mechanical properties of the cement clinker. When the added amount was 15%, the total heat of hydration that was released reached a maximum, and the mechanical properties were optimal. The 28-day flexural strength was 5.1 MPa, and the 28-day compressive strength was 31.2 MPa [6, 32]. Tian Lei used ceramic filter balls made of magnesium slag or red mud as an adsorption material; TiO_2 was loaded on the two types of ceramic filter balls to modify the surface, and then the mechanism when the TiO_2-loaded ceramic filter ball was used to remove arsenic in water was studied [33]. Zhou Shaopeng used Shanxi Yangquan bauxite, pyrolusite powder, and magnesium slag as raw materials to prepare cylindrical ceramsite proppant materials and studied the pelletizing and sintering processes of ceramsite proppant and its relevant properties [34].

3.3.3 Use Magnesium Slag to Replace Lime as a Slagging Agent in Steelmaking

The international scientific and technological cooperation project "Collaborative Research on Comprehensive Treatment and Recycling Technology for Mg slag" was undertaken by the North Minzu University, the Iron and Steel Research Institute, and enterprises in Ningxia province. Rare earth oxides or borate were used as a mineralizer in magnesium reduction to replace calcium fluoride, and fluorine-free magnesium slag was obtained. The team of co-operation used the fluorine-free magnesium slag to partially replace limestone as a slagging agent in steelmaking. Na Xianzhao et al. of the Iron and Steel Research Institute first conducted a small laboratory test to confirm that there were no adverse effects on the quality of the simulated steelmaking product, and then, they carried out two batches of multi-furnace production tests at the Ningxia steelmaking plant. The maximum content of fluorine-free magnesium slag that is used to replace lime can be as high as 30%, provided that there are no adverse effects on the quality of the steel product [6].

3.3.4 Preparation of Ca-Mg Composite Fertilizer Using Magnesium Slag

Li Yongling of Shanxi University [35] studied the preparation of magnesium slag-based slow-release Si-Ca-K composite fertilizer. The study revealed that there are few soluble components in magnesium slag. The contents of heavy metals in magnesium slag meet the national standard for organic-inorganic composite fertilizer, and the specific activity of radionuclides, the internal exposure index, and the external exposure index all meet the radioactivity requirements of the national standard for Class A decoration materials. The effective leaching amount of heavy metals in magnesium slag is far below the minimum limit of toxicity identification. Also, magnesium slag contains elements that are necessary or beneficial for soil and crops, indicating that it has a high environmental safety and poses a low pollution risk. Magnesium slag can be modified with potassium carbonate to prepare a new type of slow-release Si-Ca-K composite fertilizer. A corn potting experiment used the fertilizer and verified that it improved the agronomic characteristics of corn and increased the amount of dry matter and yield of corn. Additionally, acidic soil is more conducive to the effective release of Si and K in composite fertilizer than alkaline soil. Ge Tian [36] studied the fertility characteristics and agricultural environmental risk assessment of Si-K fertilizer that is made from magnesium slag. Liang Yiran and Mao Jia [37, 38] of Taiyuan University of Technology also studied the feasibility and technological process of preparing Si-Ca-K composite fertilizer from magnesium slag. Wang Yan of Zhengzhou University [39] studied the comprehensive utilization of magnesium slag to prepare Si-Ca-Mg fertilizer. The results indicated that the amounts of harmful trace elements in magnesium slag were not high and were all lower than the requirements in "GB8173-87" for Pb and Cd. The prepared Si-Ca-Mg fertilizer contained 20.34% of effective silicon; 85% of the fertilizer particles can pass through a 60-mesh standard sieve, and the moisture content was 0.89%, which meets the national standard for silicon fertilizer.

3.4 Advantages and Technical Difficulties in Recycling Utilization of Magnesium Slag

For the magnesium slag produced by silicothermal reduction via the Pidgeon process, the main chemical composition is 40–50% calcium oxide and 20–30% silicon dioxide, in addition to unreacted magnesium oxide and iron oxide. Heavy metals are rarely detected. Therefore, the recycling process can more easily be conducted without the need to detoxify heavy metals, and this is a major advantage of the comprehensive use of magnesium slag. However, there are several technical problems to solve: (1) The particles of magnesium slag are small, easily cause dust to rise, are light in specific gravity, and are inconvenient to transport. Thus, there is an effective radius for the economic feasibility for the bulk use of magnesium slag. Meanwhile, magnesium smelters are often established in remote areas, where raw material mines are located. They are far away from economically developed and industry-intense areas, and this makes recycling magnesium slag difficult. Although it is technically possible to change the proportion and appearance of magnesium slag through densification and modification, the economic feasibility must first be considered for the bulk use of magnesium slag, and the increase in processing costs often outweighs the gains. (2) In addition to landfilling, the current magnesium slag use in China mostly involves replacing limestone to prepare cement clinker. However, as 3–8% MgO remains in the magnesium slag, there is a later expansion in the long-term use of cement products. As a hidden danger, the later expansion causes product failure, and thus, it requires serious study and treatment. (3) Calcium fluoride in untreated magnesium slag reacts easily when is stored in the open air and is landfilled. Furthermore, it is converted at high temperature during treatment [40, 41], and there is even a risk of the spread of fluorine pollution. Therefore, it needs to be given more concentration during further recycling process.

References

1. Gao F, Nie Z, Wang Z et al (2008) Environmental assessment of energy usage strategies for magnesium production using the Pidgeon process. J Beijing Univ Technol 34(6):646–651
2. Wu L, Han F, Yang Q et al (2012) Fluoride emissions from Pidgeon process for magnesium production. In: Paper presented at the 27th international conference on solid waste technology and management, Philadelphia, 11–14 March 2012
3. Han F, Yang Q et al (2012) Environmental performance of fluorite used to catalyze MgO reduction in Pidgeon process. Adv Mater Res 577:31–38
4. Zhang Y (2013) Researching on the dynamics and parameters optimization of smelting magnesium by silicon-thermo-reduction. Dissertation, Jilin University
5. Ramakrishnan S, Koltun P (2004) Global warming impact of the magnesium produceng the Pidgeon process. Resour Conserv Recycl 42:49–64
6. Han F, Wu L (2017) Recycling utilization of industrial solid waste. Science Press, Beijing
7. China Science Research (2019) Development status and investment strategy report of Chinese magnesium industry (2018–2023)

8. Fan B, Duan L et al (2015) Fractal characteristics on Pore structure of Magnesium slag as desulfurizer. J Eng Thermophys 36:678–682
9. Fan B, Yang J et al (2013) Study of the denaturing of the agent for desulfurizing magnesium slag through hydration by using additives. J Eng Therm Energ Power 28:415–419+440–441
10. Han F, Jia L et al (2019) Effect of crystal structure on the desulfurization reactivity of magnesium slag. Chem Ind Eng Prog 38(07):3319–3325
11. Chen J, Han F et al (2019) Study on dissolution characteristics and desulfurization performance of quenching hydration magnesium slag. Boiler Technol 50(02):1–5+34
12. Ji K, Hou Y et al (2016) Experiment on desulfurization performance of magnesium slag modified by quenching hydration. Chin J Environ Eng 10(12):7235–7240
13. Duan L (2015) Study on fractal characteristics of hydration products from quenching magnesium slag. Dissertation, Taiyuan University of Technology
14. Feng L (2018) Study on wet desulfurization of various forms magnesium slag. Dissertation, Taiyuan University of Technology
15. Wang X (2010) Experimental investigation on reaction characteristics for magnesium residues as sorbent. Dissertation, Taiyuan University of Technology
16. Zhao J (2011) A study on dissolution mechanism and desulfurization performances of Alkaline industrial wastes. Dissertation, Shandong University
17. Xiao Y, Gao Y (2018) Preparation and performance test of new environment-friendly desulfurizer containing magnesium slag. Environ Eng 36(11):133–136
18. Xu X (2011) Investigation of magnesium slag via Pidgeon Process as Sulfur fixation agent for coal burning. Dissertation, Jiangxi University of Technology
19. Xu X, Chen Y et al (2017) Industrial test of desulfurization of hot metal by magnesium slag. Light Metals 1:42–44
20. Cui Z, Li S et al (2016) Drying shrinkage characteristics of magnesium slag composite admixture concrete. J Sichuan Univ (Engineering Science Edition) 48(02):207–212
21. Li Z (2016) Experiment study on carbonization properties of magnesium slag concrete. Dissertation, Ningxia University
22. Cui Z, Zhou K et al (2013) Study on the strength characteristics of magnesium slag fine aggregate concrete. Concrete 06:67–69
23. Chen G (2015) Study on controllable expansion of cementitious materials for preparing magnesium slag. Dissertation, Xi'an University of Architecture and Technology
24. Ji Y, Li Y et al (2017) Properties of quenching magnesium slag cement cementitious materials. J Xi'an Univ Arch Tech 49(02):277–283
25. Peng X, Wang K et al (2013) Hydraulic potential stimulation and bricks preparation of magnesium slag. J Chongqing Univ 36(03):48–52+58
26. Ji G (2016) Research on preparation and performance of non-autoclaved curing aerated concrete prepared from alkali activated magnesium slag. Dissertation, Chongqing University
27. Luo F (2010) Study and application of minor clinker magnesium slag cementitious material. Dissertation, Jilin architectural and civil engineering institute
28. Xiao L, Luo F et al (2009) The analysis on mechanism of using magnesium slag to prepare the cementitious material. J Jilin Inst Arch Civil Eng 26(05):1–5
29. Xiao L, Luo F et al (2011) Study of new magnesium slag energy-saving insulation wall material. Xxxxxx 38(07):21–23
30. Xiao L, Wang S et al (2008) Status research and applications of magnesium slag. J Jilin Inst Arch Civil Eng 01:1–7
31. Fang R, Che S et al (2014) Study on properties of cement clinker calcined with magnesium slag as one of raw materials. Cement 11:26–28
32. Zhao S, Han F et al (2017) Preparation of composite slag sulphoaluminate cement clinker from electrolytic manganese-magnesium. Bull Chin Ceram Soc 36(05):1766–1772+1776
33. Tian L (2010) Study on magnesium slag and red mud utilization of ceramic filter ball to remove As from wastewater. Dissertation, Wuhan University of Technology
34. Zhou S (2014). Effect of magnesium slag doping on the property of ceramsite proppant material. Dissertation, Taiyuan University of Science and Technology

35. Li Y (2016) Preparation and properties of slow releasing Silicon-potash fertilizer by using magnesium slag. Dissertation, Shanxi University
36. Ge T (2016) Fertility properties of magnesium slag-based Si-K fertilizer and agricultural environmental risk assessment. Dissertation, Shanxi University
37. Liang Y (2015) Feasibility of using magnesium slag to produce Si-Ca-Mg compound fertilizer and research of improving the content of available silicon. Dissertation, Taiyuan University of Technology
38. Mao J (2016) Using magnesium reducing slag to produce Ca-Mg-Si compound fertilizer and its applied basic research. Dissertation, Taiyuan University of Technology
39. Wang Y (2012) Comprehensive utilization of magnesium slag study and analysis on preparation of silicon, calcium and magnesium fertilizers. Dissertation, Zhengzhou University
40. Han F et al (2012) Fluoride evaporation during thermal treatment of waste slag from Mg production using Pidgeon process. Adv Mater Res 581–582:1044–1049
41. Wu L et al (2013) Fluorine vaporization and Leaching from Mg Slag treated at different conditions. Adv Mater Res 753–755:88–94

Chapter 4
Decontamination of Magnesium Slag

Abstract Magnesium slag is a valuable resource when proper disposal and utilization are adopted. The main pollution of magnesium slag caused are dust pollution and fluorine pollution. Formation mechanism of fine magnesium slag dust is investicated. Alike steel slag, phase transition of dicalcium silicate in magnesium slag around 500 °C brings volume expansion, which causes the disintegration and powdering of magnesium slag. Chamical stabilization experiments of the slag were carried out. The effects of phosphorus compounds, boron compounds and rare earth oxides doping on stabilization of magnesium slag was discussed. Catalytic action of calcium fluoride on magnesium smelting was revealed. Fluoride-free mineralizers for magnesium smelting were tested in industrial scale.

Keywords Magnesium slag · Dust pollution · Volume expansion · Chamical stabilization · Phase transition

4.1 Fixation of Magnesium Slag Dust

4.1.1 Formation Mechanism of Fine Magnesium Slag Dust

When the Pidgeon process is used to smelt metallic magnesium, crude magnesium ingots are removed after the reduction reaction is complete; meanwhile, magnesium slag is discharged from the reduction tank. The magnesium slag is transported to a slag yard and then cooled naturally to room temperature. The temperature of magnesium smelting is 1200 °C, and so, magnesium slag is also at this temperature before it leaves the reduction retort. When hot pellet-shaped magnesium slag is discharged and transported to a slag yard, it becomes a fine powdery magnesium slag after cooling and weathering. Sometimes, to accelerate the cooling and to increase the turnover rate of a slag yard, water is poured on the hot magnesium slag pile to accelerate the cooling process.

In studying the phase transformation of dicalcium silicate (Ca_2SiO_4, C_2S) in steel slag, scientists discovered that C_2S has five crystal forms, namely α-, α'$_H$-, α'$_L$-, β-, and γ-C_2S. Among these phases, the β phase is a monoclinic system that is a stable phase at high temperature and a metastable phase at normal temperature; the γ phase

L. Wu et al., *Comprehensive Utilization of Magnesium Slag by Pidgeon Process*, SpringerBriefs in Materials, https://doi.org/10.1007/978-981-16-2171-0_4

is an orthorhombic system that is a stable phase at normal temperature. When the temperature changes from high to low, C_2S undergoes several phase transitions; one involves a 12–14% expansion in volume when the β-C_2S phase converts to the γ-C_2S phase in the range of 675–490 °C [1–3]. The phase transition diagram of C_2S is shown in Fig. 4.1 [1]. The phase transition phenomenon of magnesium slag is similar to that of steel slag. CaO and SiO_2 accounting for 70–90% of magnesium slag and can form β-C_2S at a smelting temperature of 1200 °C inside a reduction retort. During a slow natural cooling process, β-C_2S transforms into γ-C_2S in the range of 500–700 °C, and this is accompanied by a volume expansion. Such a volume expansion causes the disintegration and powdering of magnesium slag, and this causes pellet-shaped magnesium slag to turn into powder.

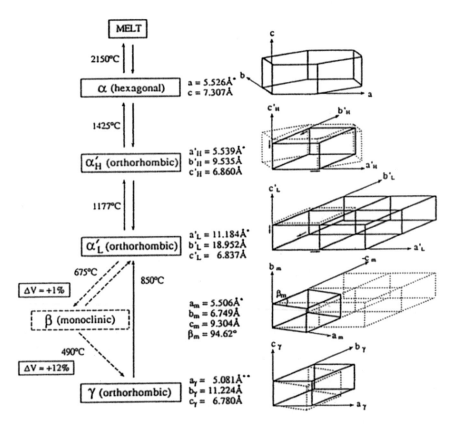

Fig. 4.1 Phase transition diagram of C_2S (Reprinted from Ref. [1] Copyright 1992, with permission from John Wiley and Sons)

4.1.2 Volume Stability of Magnesium Slag

The volume stability of magnesium slag is an important factor that affects the recycling and utilization of magnesium slag. If building materials that contain magnesium slag expand and crack in a later stage, it will seriously affect the use of the materials. Therefore, similar to that of steel, the volume stability problem of magnesium slag has attracted the attention of researchers at home and abroad. Before the phase transition, magnesium slag is in the shape of egg-sized pellets and has certain strength, which is convenient to collect and transport. It is also convenient to use magnesium slag instead of limestone as a slagging agent in steel-making. However, all magnesium slag in its stock yard is in the form of powder. The powdery slag has a light bulk density and small particle size, and thus, it easily causes dust flying, polluting the environment and making recycling and use inconvenient. On the basis of research results on the stability of steel slag, the volume stabilization of magnesium slag can be achieved through two approaches: a physical method and a chemical method.

4.1.2.1 Physical Stabilization of C_2S

The physical method to stabilize C_2S in magnesium slag is the rapid cooling of β-C_2S via air cooling or water cooling to inhibit the transition into the γ-C_2S phase and to prevent the formation of fine powder. Chan et al. prepared pure C_2S samples and then conducted a study on the stability of C_2S during the $\beta \rightarrow \gamma$ phase transition under different cooling conditions [3]. They concluded that the inhibition of the phase transition from β-C_2S to γ-C_2S is related to the phase transition history, the driving force during cooling, and the amorphous phase of the dicalcium silicate sample. They also proposed a hypothesis regarding the critical particle size for the phase transition. Yang Qingxing et al. from Lulea' University of Science and Technology in Sweden conducted research on the physical method of stabilizing steel slag and effectively achieved the physical stabilization of steel slag under laboratory conditions [4, 5]. Jiang et al. of Anhui University of Technology studied aging changes of steel slag during a 20-day period after a rapid air cooling process [6]. Zhu of Nanjing Forestry University immersed steel slag in room-temperature water to reduce the volume expansion of steel slag and conducted a comparative experiment to study the strength of a steel slag mixture before and after the immersion treatment [7]. Professor Cui of Ningxia University conducted an experiment on the expansibility of magnesium slag and explored the expansion mechanism [8]. The author of this book also conducted air cooling and water cooling experiments on magnesium slag generated via the Pidgeon process and observed that the stabilizing effects of both cooling processes on the magnesium slag are obvious. Whether for steel slag or for magnesium slag, it is easy to stabilize C_2S through physical method under experimental conditions. However, on industrial production sites, energy consumption and site conditions must be considered, and it are normally difficult to meet the requirements of the continuous and stable treatment of large quantities of slags.

4.1.2.2   Chemical Stabilization of C_2S

Chemical stabilization of C_2S is accomplished by adding a small amount of chemical agents that contains certain ions to form a solid solution. The solid solution enters into C_2S grains or is present at the grain boundaries, and this inhibits the phase transition from β-C_2S to γ-C_2S, thereby avoiding the powdering of magnesium slag caused by volume expansion; this reduces the damage that dust pollution causes to the environment. Kawasaki Iron and Steel Company of Japan used chemicals containing P_2O_5 and B_2O_3 to study the stability of stainless steel slag and found that following methods can be used to stabilize β-C_2S: (1) The original Si^{4+} or Ca^{2+} ions are substituted with ions that have a radius smaller than that of Si^{4+} or larger than that Ca^{2+}. (2) Ions with a C/R (valence/ion radius) ratio < 2 or > 9.5 are used to substitute existing ions [2], as follows.

$$B^{3+} \quad P^{5+} \quad Si^{4+}$$
$$0.22 \quad 0.33 \quad < 0.4(\text{Å})$$
$$Ba^{3+} \quad Sr^{2+} \quad Ca^{2+}$$
$$1.36 > 1.16 > 0.99(\text{Å})$$
$$1.36 > 1.16 > 0.99(\text{Å})$$

In 2000 at the 6th international conference on molten slags, fluxes, and salts, German scientist Jürgen Geiseler introduced a series of elemental ions that can stabilize C_2S, as shown in Fig. 4.2 [9].

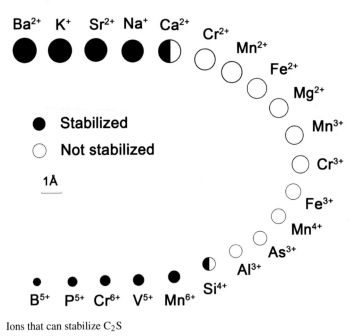

Fig. 4.2 Ions that can stabilize C_2S

The author of this book added a mineralizer that contained P and B ions to magnesium slag. Experimental results confirmed that the added mineralizer stabilized magnesium slag. The experiments are described in detail in the following sections [10, 11]. Feng [12], Zhang [13], Huang [14], and other researchers have done similar research on steel slag and have drawn similar conclusions.

4.1.3 Experimental Study on Inhibition of C_2S Phase Transition

This section introduces experimental research of the author's research group regarding the inhibiting of phase transitions of C_2S in magnesium slag, thereby reducing the degree of pulverization of magnesium slag.

4.1.3.1 Phosphorus-Doping Experiment

To stabilize of C_2S in magnesium slag, make magnesium slag be lumpy and easy to collect, store, and transport. The lumpy magnesium slag diminishes the dust pollution caused by powdering of the slag and increases the proportion of $\beta\text{-}C_2S$ with gelling activity in the slag, which is favorable for using magnesium slag in construction materials [10]. Commercial phosphates were used as stabilizers for C_2S, and stabilization tests of magnesium slag were carried out.

Experimental material: The magnesium slag used in the experiments came from Fugu District Plant of Huiye Magnesium Group Co., Ltd. in Ningxia province. The main composition of the slag is shown in Table 4.1.

The phosphate stabilizers used in the experiment include fluorapatite, containing 17% phosphorus (industrial grade, Sihui Feilaifeng Non-metallic Mineral Material Co., Ltd.); calcium dihydrogen phosphate containing 22% phosphorus (food-grade, Lianyungang Xidu Biochemical Co., Ltd.); and calcium dihydrogen phosphate containing 22% phosphorus (feed-grade, Mianyang Shenlong Feed Co., Ltd.). Similar products produced in Sweden were tested for comparison. The added amount of stabilizer ranged from 0.5 to 8 wt%.

Main equipment: An X-ray diffractometer (Shimadzu XRD-6000), a laser particle size distribution analyzer (Honeywell Microtrac X-100), and an ultraviolet spectrophotometer (Beijing General TU-1810) were used.

Experimental procedures: Magnesium slag was placed in a vibrating mill and ground for 3 min; then, it was passed through a 40-mesh sieve, and kept for later use.

Table 4.1 Composition of the magnesium slag sample used in the phosphorous experiment

Composition	MgO	CaO	SiO_2	P_2O_5	Fe	Al	Mn	Na	CaO/SiO_2
%	5.12	42.35	26.78	0.061	3.85	0.604	0.061	0.979	1.58

The three types of phosphate reagents were added separately to the magnesium slag and stirred well. The uniform mixture was poured into a steel mold of 20 × 20 mm and pressed under 400 MPa using a hydraulic presser. After pressing, the samples were placed in a muffle furnace and calcined. The temperature was increased to 1200 °C and kept for 6 h. Then, the power was turned off, and the samples were subjected to furnace cooling. The phenomenon was observed after the samples were removed from the furnace. The amounts of stabilizer that were added to the slag were 1, 2.5, and 5% for the first group of samples and 0.5, 2, 4, and 8% for the second group of samples.

Analysis of results: (1) Morphology observation: The appearance of the first group of sintered samples is shown in Fig. 4.3a and that of the second group of samples is shown in Fig. 4.3b. In the figure, the blank sample is the reference magnesium slag sample without stabilizer. The reference samples were severely powdered; in contrast, all of the samples with stabilizers remained intact, and they are still not powdered after being stored for one month.

(2) XRD analysis: The XRD patterns of stabilized samples are shown for comparison in Fig. 4.4, where (a) is the reference sample without an added phosphorous compound, (b) is magnesium slag with 5 wt% of food-grade calcium dihydrogen phosphate, (c) is magnesium slag with 5% feed-grade calcium dihydrogen phosphate, and (d) is magnesium slag with 5% fluorapatite.

As seen in the XRD patterns, the diffraction peaks of the γ-C_2S phase in the phosphorus-stabilized sample are reduced significantly compared to the diffraction peaks of the reference sample. The strong diffraction peaks near $2\theta = 30°$ disappear. The results indicate that in phosphorus-stabilized samples, γ-C_2S decreased and β-C_2S increased compared to the reference sample without stabilized agents. Also, the

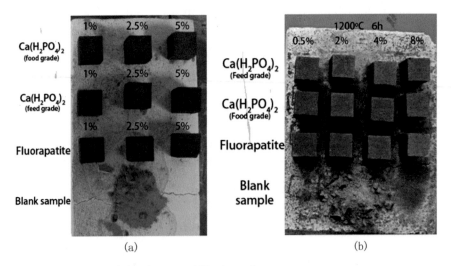

Fig. 4.3 Morphology of phosphorous-stabilized samples

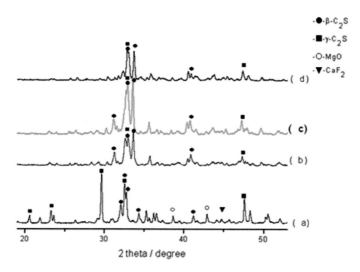

Fig. 4.4 XRD patterns of phosphorous-stabilized samples

three phosphorus-containing stabilizers do not show significant differences in the stabilizing effect.

Within a temperature range of 900–1200 °C, Fig. 4.5 shows the existential state of phosphorus simulated by Factage 6.2 using the compositions of the phosphorus-stabilized sample as the input data for the software. As seen from the figure, at high temperature, the added phosphorus acts in the form of a phosphate radical and forms a solid compound with calcium (or fluorine).

4.1.3.2 Boron-Doping Experiment

Using boron-containing compounds as stabilizers to modify magnesium slag can suppress powdering and dusting of magnesium slag. This reduces the harm that magnesium slag can cause to the environment and improve the environmental conditions of a slag field and its surroundings [15, 16].

Experimental materials: The magnesium slag used in the experiment was from the Fugu Plant of Ningxia Huiye Magnesium Co., Ltd. Three types of borate reagents were used as stabilizers to investigate the effect that borate reagents have on preventing magnesium slag from powdering. Table 4.2 lists the compositions and melting points of three boron-containing compounds: anhydrous borax ($B_4Na_2O_7$), G-Vitribore 25, and boric acid (H_3BO_3). These three types of boron-containing compounds are all commercially available chemical reagents and are referred to as DB, GB, and H1, respectively. At higher temperature, the B_2O_3 and Na_2O in DB and GB also have a certain stabilizing effect on polymorphic C_2S. These two boron-containing compounds are also reagents that are commonly used as dusting inhibitors of steel slag in many plants. Before use, all three of these boron-containing

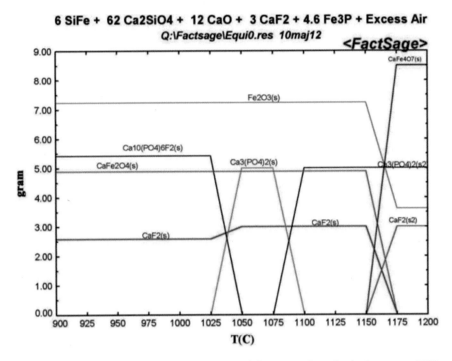

Fig. 4.5 Behavior predictions of phosphorous-containing magnesium slag in the range of 900–1200°C

Table 4.2 Contents (wt%) and melting points (°C) of boron-containing compounds

Borate	CaO	SiO$_2$	B$_2$O$_3$	Na$_2$O	MgO	P$_2$O$_5$	Melting point
DB			69	30.8			742
GB	8.8	29.4	23.5	23.5	0.3	1.4	696
H1			56.5				

compounds were ground and sieved as described above. Table 4.2 shows the main compositions and melting points of these boron-containing compounds.

Preparation of samples

Magnesium slag must be dried before it is used; 30 mg of dried magnesium slag was mixed with each of the above three boron-containing compounds. The mass fraction of the added boron-containing compound was 0–1 wt% of the magnesium slag. A hydraulic presser was used to press the mixture into a square block with dimensions of 40 × 40 × 6 mm. The pressed compact blocks were placed in a muffle furnace, and the temperature was increased to 1200 °C. The samples were kept at this temperature for 2–6 h to carry out the sintering. The sintered blocks were then furnace-cooled. The samples were removed, and the cooled magnesium slag samples were examined to evaluate the effects of the boron-containing stabilizers.

Morphologies and phases of samples

Figure 4.6 shows the morphology of the first group of samples, which were sintered at 1200 °C and held for 5 h. Among the samples, sample (a) is the reference sample without stabilizer, (b) is the sample with 0.53 wt% DB, and (c) is the sample with 0.54 wt% GB. Figure 4.7 shows the morphology of the second group of test samples, which were sintered at 1200 °C and held for 6 h. In Fig. 4.8, a is the magnesium slag with 0.59 wt% boric acid, (b) is the magnesium slag with 0.95 wt% boric acid, (c) is the magnesium slag sample with 0.34 wt% DB, and (d) is the reference sample without stabilizer. The results of the phosphorus-stabilized experiment are similar to those of the magnesium slag sample without stabilizer, which were severely powdered after cooling. Specifically, the test piece disintegrated and collapsed into fine powders. However, the sample with a boron-containing stabilizer basically retained the original morphology. A comparison of the different stabilizers indicates that the stabilizing effect of DB (anhydrous borax) is not as good as those of GB, G-Vitribore 25, and boric acid. In particular, in the second group of experiments, damage occurred in the sample with the DB stabilizer after a longer of holding time.

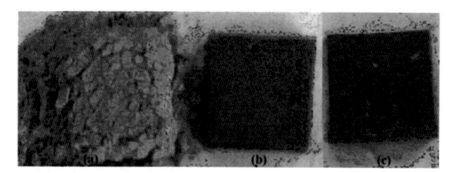

Fig. 4.6 Morphologies of boron-stabilized slag samples in group 1: **a** reference sample without stabilizer, **b** sample with 0.53 wt% DB, and **c** sample with 0.54 wt% GB

Fig. 4.7 Morphologies of boron-stabilized slag samples in group 2: **a** sample with 0.59 wt% of boric acid, **b** sample with 0.95 wt% of boric acid, **c** sample with 0.34 wt% DB, and **d** reference sample without stabilizer

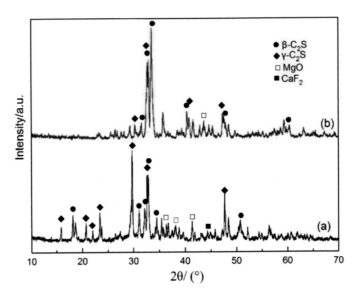

Fig. 4.8 XRD patterns of boron-stabilized samples: **a** reference sample and **b** sample with 0.53 wt% DB

The phase analysis of the original magnesium slag, which served as a reference sample, and of the sintered magnesium slag with a boron-containing compound was carried out using an X-ray diffractometer (Shimadzu X-6000). The XRD patterns are shown in Fig. 4.8.

In Fig. 4.8, (a) is the reference sample without a stabilizer and (b) is the sintered magnesium slag sample with 0.53 wt% DB. As seen from the XRD pattern of the reference sample in Fig. 4.8a, γ-C$_2$S is the main phase, and it is the main reason that the powdering phenomenon of magnesium slag occurs. β-C$_2$S is detected as a secondary phase, and this indicates that a small amount of β-C$_2$S is present in the powdered magnesium slag. Magnesium oxide is also present. A trace amount of CaF$_2$ is also detected, and it only accounts for 2.5–3 wt% in the raw materials. As seen in Fig. 4.8b, the content of γ-C$_2$S in the magnesium slag sample with 0.53 wt% DB after sintering was significantly lower than that in the reference sample. Also, C$_2$S was mainly present in the form of the β-C$_2$S phase, indicating that the added boron ions inhibited the phase transition from β-C$_2$S to γ-C$_2$S in the sample. Interestingly, the diffraction peaks of MgO were also weakened, and this is possibly because some of the free MgO in the slag was stabilized in the stationary phase. The CaF$_2$ phase was not detected.

The results of the boron-stabilized samples are similar to those of the phosphorus-stabilized samples. Compared with the reference sample, the diffraction peaks of the γ-C$_2$S phase in the boron-stabilized sample were significantly reduced; the strongest γ-C$_2$S diffraction peak at a 2θ angle near 30° was much weaker, and the intensity was significantly lower.

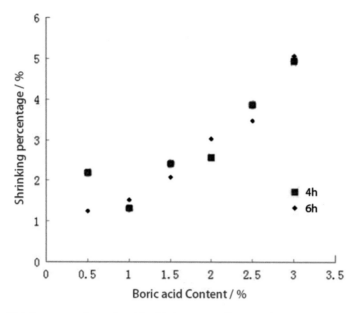

Fig. 4.9 Shrinkage rates of samples with added amounts of borate acid

Shrinkage and influencing factors

The volume shrinkage of samples with different contents of boric acid was measured after sintering. The volume shrinkage rate is equal to the value of (volume of original sample minus volume of sintered sample)/volume of original sample. The changes in shrinkage with respect to mineralizer content are shown in Fig. 4.9. The data represented by red squares in the figure are the shrinkage of the sample after 4 h of temperature holding time, and the data represented by black diamonds are the shrinkage of the sample after 6 h of temperature holding time.

As seen in Fig. 4.9, the overall shrinkage of the sintered samples increase linearly with an increase in the amount of added boric acid. There are no essential differences between the results for the samples that had temperature holding times of 4 and 6 h. In particular, the shrinkage of samples with different temperature holding times is almost consistent when the amount of added boric acid reached a maximum value of 3%.

4.1.3.3 Doping Experiment Using Rare Earth Oxides

In the doping experiment, the rare earth oxides La_2O_3, Y_2O_3, and Ce_2O_3 (purchased from Baotou Rare Earth Research Institute) were used as stabilizers. The percent range of the added rare earth oxide stabilizer was 0.5–5%. The experimental temperatures were 1100, 1150, and 1200 °C, and the holding time range was 0.5–6 h. The other experimental conditions were the same as those in previous phosphorus-doping and boron-doping stabilization experiments. Figures 4.10, 4.11 and 4.12 show photos from Ce_2O_3-doping (or La_2O_3-doping) stabilization experiments.

Figure 4.10 shows photos of samples with Ce_2O_3 used as a stabilizer. The sample was held at temperatures of 1200, 1150, and 1100 °C for 6 h, and the amount of added stabilizer gradually increased (0.5–5%) in the samples that are shown from left to right. As seen in the figure, a large percentage of stabilizer is beneficial for volume

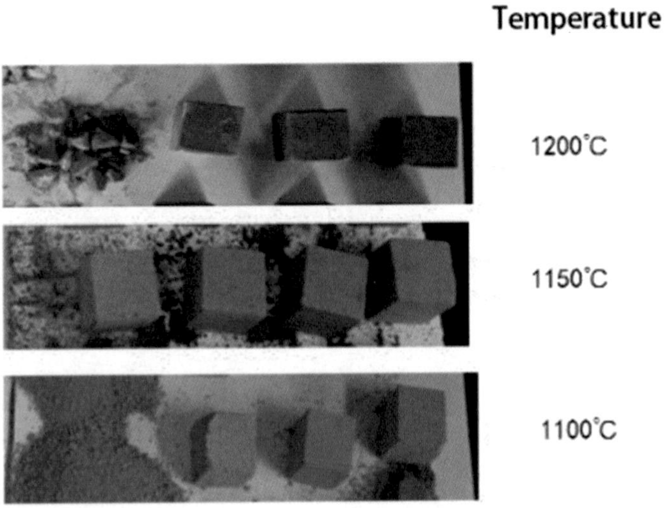

Temperature

1200°C

1150°C

1100°C

Fig. 4.10 Stabilizing effects of Ce_2O_3-doping at different temperatures

Fig. 4.11 Stabilizing effects of Ce_2O_3-doping

Fig. 4.12 Experimental effects of the La₂O₃ doping experiment

stability of magnesium slag; the sample that was held at a temperature of 1200 °C had the best compactness and the largest volume shrinkage. For comparison, Fig. 4.11 shows photos of samples that had different amounts of added CeO_2 with a holding time of 5 h and a temperature of 1200 °C. Compared to the results of samples that had a holding time of 6 h, there are no significant differences. As seen from several sets of experimental photos, the sample with the minimum added amount of 0.5% obviously did not meet the requirements for stabilizing magnesium slag.

Figure 4.12 shows photos of samples that used La_2O_3 as a stabilizer. The samples were held at 1200 °C for 5 h, and the amount of added La_2O_3 increased continuously in the samples ordered#1–#9, which had increasing content from 0.5–3 wt%. It is clear from Fig. 4.12 that La_2O_3 has a remarkable stabilizing effect. However, because of the costs of practical applications, the role that rare earth oxides play in stabilizing magnesium slag has not been further studied

4.2 Reduction of Fluorine in Magnesium Slag

4.2.1 Catalytic Effects of Calcium Fluoride on Reduction of Mg

The traditional silicothermic process of magnesium smelting via Pidgeon process is to reduce dolomite using ferrosilicon at 1200 °C under negative pressure. Calcium fluoride acts as a catalyst in the silicothermic reduction and does not participate in the reaction. Therefore, after the reaction is complete, most of the fluorine-containing components remain in the slag.

The chemical reaction of smelting magnesium via the Pidgeon process is as follows:

$$2CaO \cdot MgO(s) + (Fe)Si(s) \rightarrow 2Mg(g) + Ca_2SiO_4(s) + Fe(s)$$

The main raw materials that are used in magnesium smelting include calcined dolomite ($CaO \cdot MgO$), ferrosilicon (Fe)Si, and fluorite (CaF_2). In industry, the proportion of raw materials is usually as follows:

calcined dolomite: ferrosilicon: fluorite = 100:20:3

The converted mass ratio is as follows: 81.3:16.3:2.44 (wt%)

Barua et al. [17] used the simulated experimental device shown in Fig. 4.13, prepared a pellet of magnesium smelting raw material, and hung it on a quartz wire to detect the weight loss of the pellet during the reaction. The temperature inside the reactor was controlled to be between 1070-1250 °C. H_2 gas was introduced into the reactor with a flow rate of 6 l/min, so that the Mg vapor that is generated via the reduction reaction can leave the reactor with H_2 gas when a flow of H_2 gas is passed through the pellets. As a result, the Mg vapor pressure on the surface of the pellet is close to the pressure of vacuum magnesium smelting, and this meets the

Fig. 4.13 Equipment used in the kinetically simulated magnesium smelting experiment (Reprinted from Ref. [17] Copyright 1981, with permission from Taylor & Francis Group)

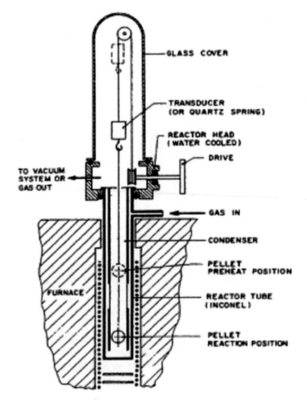

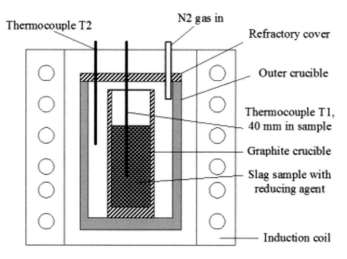

Fig. 4.14 Reduction experiment equipment used in smelting magnesium

thermodynamic conditions of vacuum magnesium smelting. In addition to measuring the weight loss of the pellet in the test, Barua et al. also analyzed and characterized the pellet after the test. They used theoretical calculations and obtained research results on the kinetics conditions for magnesium smelting. The author of this book used a device, as shown in Fig. 4.14, to carry out silicothermic reduction of MgO using CaF_2 as a catalyst. In the magnesium smelting simulation experiment, a nozzle was used to blow N_2 directly into a small crucible to reduce the partial pressure of magnesium vapor on the surface of the raw material in magnesium smelting. As shown in the figure, the thermocouple T1 was buried in the center part of the experimental sample to measure the temperature inside the sample. The thermocouple T2 was fixed in the furnace cavity to measure the furnace temperature.

Figure 4.15 shows the curves of the sample temperature and furnace temperature that were measured with respect to time under experimental conditions. The experimental raw material had a weight of 350 g, and the proportion was according to the industrial standard (including fluorite) for magnesium smelting. The temperature inside the sample (T_1) is marked in red, and the temperature inside the furnace (T_2) is marked in blue. As seen in Fig. 4.16, during the first 50 min of the simulation experiment, T_1 was always lower than T_2. T_1 began to increase rapidly after 45 min, caught up with T_2 at about 50 min and remains close to T_2 until the end of the experiment. To study the effect that the mineralizer has on the temperature inside the material used in the simulated magnesium smelting experiment, the rate of temperature increase inside the material (i.e., the heating rate) was calculated using the following Formula:

$$\text{Heating rate} = \Delta T / \Delta t$$

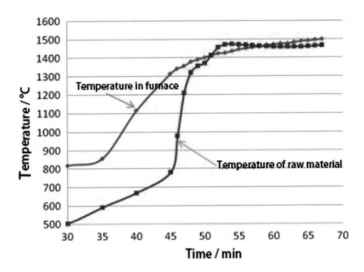

Fig. 4.15 Changes in temperature inside the material (T1) and inside the furnace (T2) with respect to time

where ΔT denotes the temperature difference between two adjacent points of measurement, and Δt is the time interval of the temperature measurement.

For the simulation tests with and without CaF_2, the changes in the material's heating rate relative to the furnace temperature are shown in Fig. 4.16. In the

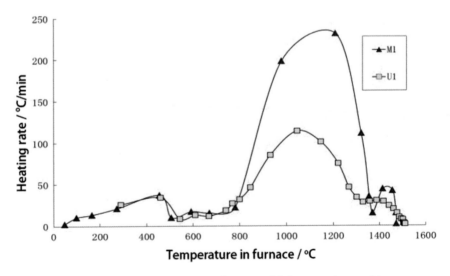

Fig. 4.16 Effects of CaF_2 on the heating rate of raw materials in magnesium smelting

figure, M1 is a standard formulation of material for magnesium smelting production containing calcium fluoride, and U1 has the same formulation as M1 but does not contain calcium fluoride.

In the low temperature stage (below 800 °C), the heating rates of the two test materials are similar. Starting from a furnace temperature above 800 °C, the heating rate of M1 observably increases, and it is close to 250 °C/min at a furnace temperature of 1200 °C. The heating rates of the raw materials are closely related to the reaction rate of magnesium smelting. As ferrosilicon melts at high temperature, a liquid phase is generated, and the flow of the liquid phase greatly improves the kinetics conditions of the reduction reaction, accelerating the reaction. Therefore, the heating rates of samples with and without CaF_2 both start to increase. However, the comparison between M1 and U1 curves clearly show that the heating rate of M1 containing CaF_2 is almost double that of U1 without CaF_2; this indicates that adding CaF_2 increases the heating rate significantly, thereby increasing the reduction rate of MgO. The experimental results confirm that the catalytic effect of CaF_2 as a mineralizer is very obvious in the reduction of magnesium oxide.

4.2.2 Magnesium Smelting Pilot Test

4.2.2.1 Equipment Used in the Pilot Test

To be as close as possible to the actual production conditions of magnesium smelting via the Pidgeon process and to optimize the formula of the raw materials, a pilot test was used to simulate magnesium smelting experiments. The reduction retort used in industry was scaled down, and two identical retorts that can each hold 10 kg of raw materials were designed and manufactured. The reduction retorts were placed side-by-side in a muffle furnace with temperature control. The furnace size of the muffle furnace was $810 \times 550 \times 375$ mm; the equipment power was 25 kw, and the heating rate was controllable. It takes about 2.5–3 h for the furnace to reach 1130 °C. Two-stage vacuum pumps were connected in series to ensure that the vacuum in the reduction retort reached 7–13 Pa, and this meets the requirements for magnesium smelting. The reduction retort was made of heat-resistant steel pipe. One end of the pipe was sealed, and other end was welded with a water- cooling jacket. To simulate industrial production conditions, an experimental device was established to compare materials with the standard formula of the Pidgeon process and the experimental formula. A schematic diagram and actual photos of the pilot experimental system are shown in Fig. 4.17. A schematic diagram of the structure of the reduction retort is shown in Fig. 4.18.

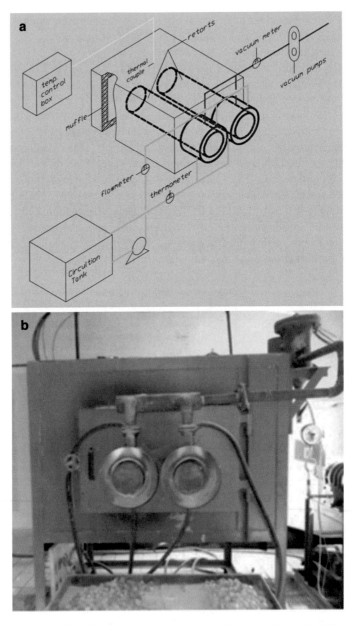

Fig. 4.17 Experimental equipment used to simulate magnesium smelting via the Pidgeon process **a** schematic diagram of the equipment's system and **b** photo of experimental equipment

Fig. 4.18 Schematic diagram of reduction retort in pilot magnesium smelting equipment 1—reduction retort 2—Mg Crystallizer 3—water jacket 4—vacuum interface 5—K/Na Crystallizer

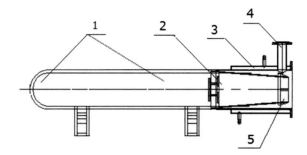

4.2.2.2 Process Flow of Pilot Magnesium Smelting Test

The pilot-scale magnesium smelting process imitates the production process of metallic magnesium, as shown in Fig. 4.19. The raw materials that were used in the simulated experiment included calcined dolomite, fluorite, and ferrosilicon, which were all obtained from Ningxia Huiye Magnesium Group Co., Ltd. The raw materials were crushed to pass through 100 mesh, mixed with different mineralizers,

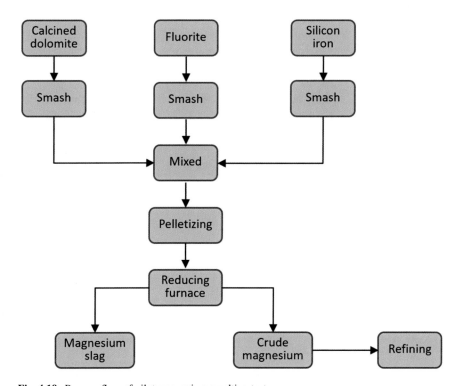

Fig. 4.19 Process flow of pilot magnesium smelting test

compressed into blocks with similar volume as the production pellets, packed into a paper bag, and placed into the retort of the pilot furnace for a comparison experiment.

The magnesium smelting process used in the experimental furnace is as follows: In the first stage, the furnace temperature was increased to 1150 °C, and then the compressed blocks of material were placed in the reduction retorts. The sample with the standard formula used in magnesium smelting production was placed in the left reduction retort (A) as the reference sample, and the sample with the experimental formula was placed in the right reduction retort (B). After the loading was complete, the mouths of the retorts were closed, and the first-stage vacuum pump was started. When the vacuum reached 80 Pa, the second-stage vacuum pump was started and run until the vacuum reached 7–13 Pa. (See the production parameters of magnesium smelting.) The current was 65 A, and the flow rate of the water in the cooling jacket was 15–20 L/h. After the reaction was complete, the power was cut off, and the furnace was cooled. Finally, the vacuum pump was turned off; the furnace mouth was opened, and the magnesium crystallizer was removed. The magnesium slag was removed and cooled to room temperature. The magnesium slag was sieved using a 40-mesh standard sieve, and then the oversize rate of the magnesium slag was calculated. The range of the reduction temperature during the entire smelting process was 1195–1205 °C, and the reduction cycle was 400 min. The inner layer of the reduction retort wall was made of heat-resistant steel and might desquamate under high temperature and oxidation conditions; small pieces of iron scraps enter the magnesium slag and may affect the accuracy of the calculated weight composition of magnesium slag.

4.2.2.3   Influence of the Pellet Formation Method on Magnesium Smelting

To ensure the complete reaction of raw materials in the reduction retort, the raw materials must be pressed into briquette-sized pellets with certain strength. If the pellets are not strong enough, they break easily during the filling and reaction process and affect the reaction. However, pellets that are too dense are not conducive to the escape of magnesium vapor, which is also inadvisable. Industrial production requires a special pelletizer to prepare pellets of the raw materials. Because many varieties and small quantity of materials were used in the experiment, steel molds were used to press the sample block. To verify the influence that the formation method has on the yield of crude magnesium, one reduction retort was filled with the pellets that were made in the factory pelletizer, whereas the other reduction retort was filled with sample blocks that had the same weight and composition and were pressed in the laboratory press. They were smelted at the same time in the pilot furnace. The respective crude magnesium yields were compared, and the results are shown in Table 4.3.

The following observations are made from Table 4.3: (1) The vacuum, cooling, and heating systems of the pilot system are normal during the blank sintering experiment. Blank sintering was carried out 10 times. (2) The contents of the left and right retorts are consistent, and the magnesium yields are almost the same; this indicates that the

Table 4.3 Comparison experiment of the sample formation method

Number of experiment	Raw materials	Crude magnesium output/kg		Note
		Left retort	Right retort	
1				Blank experiment with normal running
2	Each retort has 10 kg of raw material pellets	1.65	1.65	The materials in both retorts have the same formula. All samples were pressed and shaped by a pellet-presser of industry
3		1.85	1.85	
4		1.65	1.65	
5		1.85	1.85	
6		2.00	2.05	
7	right retort: 10 kg of pellets from magnesium plant; left retort: 10 kg of pressed blocks	1.20	1.30	The materials in the left and right retort have the same formula; the sample in the right retort is formed by a pellet-presserof industry, and the sample in the left retort is shaped in Lab
8		0.75	1.40	
9		1.45	1.45	
10		1.60	1.70	
11		1.70	1.80	

reaction consistency between the left and right retort is remarkable. (3) The left and right tanks were respectively loaded with pellets made in a magnesium plant and with laboratory-pressed blocks. When the raw material formulations were identical, the crude magnesium yield of laboratory-pressed blocks was slightly lower than that of factory pellets (the result of the 8th group was abnormal), and the error was less than 10%. The formation method of the experimental sample blocks can be used to simulate the industrial pellet formation method for raw materials used in industrialized magnesium smelting.

4.2.3   Migration of Fluorine in Magnesium Smelting

In magnesium smelting via silicothermic reduction, it is necessary to add about 3 wt% of fluorite (containing 95 wt% calcium fluoride, CaF_2) as a mineralizer of magnesium reduction. The fluorine in calcium fluoride does not react with magnesium and only serves to promote the reaction. After the reaction is complete, CaF_2 is discharged with magnesium slag. Feng et al. [18] once believed that fluorine became gaseous under the high temperature in magnesium smelting and that the gaseous fluorine was pumped out using vacuum system, escaping into the atmosphere. To verify the whereabouts of fluorine, the author's team used the pilot test equipment for magnesium smelting to investigate Gao's statement, and the results are discussed [19]. The experimental data are shown in Tables 4.4 and 4.5. The fluorine balance in Table 4.5 was calculated using the experimental data in Table 4.4.

As seen from the fluorine balance shown in Table 4.5, after the reaction was complete, most of the fluorine added to the silicothermic reaction system with raw materials can be detected in the magnesium slag. This may occur because of the following reasons: CaF_2 in the raw material pellets decomposes into gas at a temperature above 1000 °C in the front zone of the reduction retort and moves to the rear zone of the reduction retort under the driving of vacuum power. When it reaches the rear cooling zone using circulating water, it condenses into a solid as the temperature drops. Because fluorine does not participate in the reaction, it is finally discharged with the magnesium slag. Therefore, most of the fluorine is not discharged directly via the vacuum system, but it is present in magnesium slag in the form of solid fluorine-containing compounds.

Figure 4.20 shows the relationship between the calculated value of fluorine in the raw material pellets and the measured fluorine content in the magnesium slag samples. As seen from the figure, the fluorine content in magnesium slag is directly proportional to the fluorine content in the raw material.

FactSageTM (FactSage 6.2) was used to simulate and calculate the migration and change of fluorine during the magnesium smelting process [19]. Table 4.6 lists the initial formulation of the input raw materials using in the simulation calculation. (The total amount was 100 g.) The proportion of each material component refers to that in the magnesium smelted via the Pidgeon process. The temperature range was 100–1200 °C, and the pressure was 10 Pa. Figure 4.21 shows a schematic diagram of changes in components with respect to temperature changes in the FACTSAGE-simulated magnesium smelting process; as seen, some components disappear and new phases form.

On the basis of the above experimental results of simulated magnesium smelting, it can be inferred that the reduction slag of magnesium produced in the Pidgeon process contains most of the fluorine that was in the raw material fluorite. This part of fluorine is discharged with the reduction slag.

4.2.4 Influence of Fluorine on Reuse of Mg Slag

In 2009, magnesium slag was inclued into national standard as a mixed material that can be used as cement production raw material. Magnesium slag contains a large amount of the silicate mineral C_2S. In addition to replacing some of the mineral raw materials when sintering cement clinker together with other raw materials, C_2S can serve as a crystal seed during the calcination process of cement materials to

Table 4.4 Trace of fluorine in the smelting process

Expe. No.	Raw materials/kg			Products/kg			Fluorine in products/wt%		
	Calcined dolomite	Ferrosilicon	Fluorite	Mg slag	Crude Mg ingot	K/Na crystals	Mg slag	Crude Mg ingot	K/Na crystals
1	5.0	0.84	0.25	4.1	0.9	0.031	2.6	0.064	0.169
2	4.5	0.84	0.25	4.2	0.8	0.023	2.95	0.042	0.166
3	8.13	1.67	0.25	7.1	1.8	–	1.57	0.022	–
4	8.13	1.67	0.25	8.3	1.85	0.061	1.24	0.047	0.07

Table 4.5 Fluorine balance

Number of experiments		1	2	3	4
Fluorine$_{in}$/g	Raw materials	115.7	115.7	115.7	115.7
Fluorine$_{out}$/g	Mg slag	106.6	123.9*	111.5	102.9
	Crude Mg ingot	0.58	0.34	0.4	0.87
	K/Na crystals	0.05	0.04	–	0.04
	total	107.2	124.3	111.9	103.8
Fluorine$_{in}$–Fluorine$_{out}$/g		8.5	–	3.8	11.9
Fluorine$_{out}$/Fluorine$_{in}$/%		92.68	–	96.69	89.74
Fluorine$_{out}$/ in Mg slag/%/Fluorine$_{in}$		92.13	–	96.34	88.95

*Note Measurement error may cause the results of experiment 2 for fluorine in magnesium slag to be larger than the amount of input fluorine

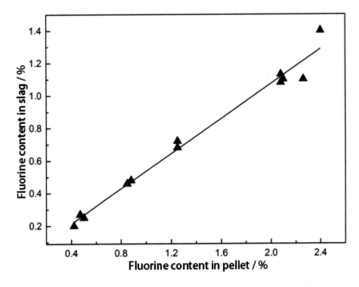

Fig. 4.20 Relationship between the percent of fluorine in the raw materials and that in the corresponding magnesium slag

Table 4.6 Input values of reactants in simulation calculations

Reactant	Si	Fe		CaO	MgO	CaF$_2$	Na$_2$O
Mass/g	12	4		56.1	25.2	2.5	1

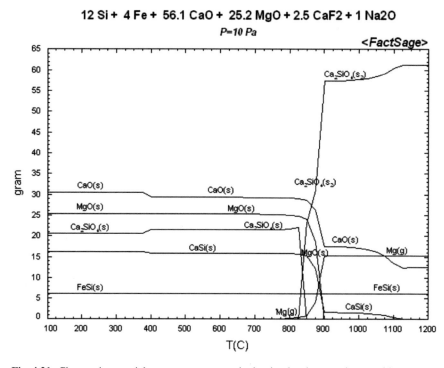

Fig. 4.21 Changes in materials versus temperature in the simulated magnesium smelting process

reduce the potential energy in crystal nucleation. Thus, magnesium slag can be used to accelerate the formation of C_3S and to promote the sintering of cement clinker. The CaO and SiO_2 that is contained in the magnesium slag can reduce the amounts of limestone and clay that are used in the raw materials of cement. This reduces the damage to the natural ecology that is caused by the excavation of the raw materials used in cement; it also reduces the energy consumption in processes during clinker calcination, such as clay dehydration and $CaCO_3$ decomposition. Therefore, using magnesium slag to prepare cement clinker has advantages that other metallurgical slags do not possess. However, under the high temperature of the sintering process of cement clinker, the CaF_2 in the magnesium slag becomes gaseous and volatilizes into the atmosphere. This therefore causes secondary pollution, which causes sufficient concern. For reuses of magnesium slag that require a high-temperature treatment, serious attention must be paid to the problem of fluorine evaporation.

Magnesium slag can replace limestone as the slagging agent used in steel-making. The results of the industrial test carried out at the Ningxia Steel Plant confirm that replacing 30% of limestone with magnesium slag as the slagging agent for steel-making was successful. However, the F components in magnesium slag might enter the molten steel, and this affects the quality of fluorine-free steel products. This requires further investigation and verification and cannot be easily ignored.

When magnesium slag is recycled and used, attention should be paid to the leaching of fluoride. The maximum fluorine emission in the National Standard of China GB5749-2006 for drinking water quality and safety is 10 mg/kg, and the leaching result of fluorine in magnesium slag far exceeds the permitted limit. If waste water is discharged during the treatment of magnesium slag, it can easily cause fluoride leaching and then pose a threat to the environment so researchers should pay special concern to it.

4.2.5   Magnesium Smelting Using Fluorine-Free Mineralizer

To solve the problems of dusting and fluorine pollution of magnesium slag, a magnesium smelting pilot test was carried out using fluorine-free mineralizer instead of CaF_2. The test is divided into two parts: (1) using boron-containing compounds to replace CaF_2 in magnesium smelting and (2) using rare earth oxides to replace CaF_2 in magnesium smelting.

Experimental materials: The experimental materials used in the pilot test were from Ningxia Huiye Magnesium Group Co., Ltd. The main raw materials were calcined dolomite, ferrosilicon, fluorite, boron-containing compounds, and rare earth oxides. The mass fraction of MgO in dolomite is 31%, the mass fraction of silicon in ferrosilicon is 75%, and the mass fraction of CaF_2 in fluorite is 95%. Industrial-grade boric acid ($H_3BO_3 \geq 99.6\%$; sulfate $\leq 0.08\%$) was selected as the boron-containing mineralizer, according to the results of the stability experiment. The rare earth oxides that were used in the experiments were cerium oxide (CeO_2), lanthanum oxide (La_2O_3), yttrium oxide (Y_2O_3), and terbium oxide (Tb_2O_3); all of rare earth oxides were purchased from Baotou Rare Earth Research Institute.

Experimental equipment: The pilot experimental equipment introduced in 4.2.2.1 was used. The sample in retort A of two parallel reduction retorts was the reference sample, the magnesium smelting raw materials were in accordance with the production formula, and fluorite was used as a mineralizer. The sample in retort B was the experimental sample, boron-containing mineralizers were used to replace some or all of the fluorite, and the other raw materials were consistent with the reference sample. The two retorts were placed under the same experimental conditions; specifically, the heating rate, magnesium smelting temperature, and degree of vacuum were completely identical.

Experiment I using a boron-containing compound to replace fluorite in magnesium smelting. Boric acid was selected as a typical boron-containing compound for the experiment. The experiment was divided into two parts: in experiment (a), all of the fluorite was replaced with boric acid, and in experiment (b), some of the fluorite was replaced with boric acid. In experiment 1(a), 8.13 kg dolomite and 1.67 kg ferrosilicon were added to each reduction retort. The amount of fluorite added to the reference group was 0.25 kg. No fluorite was added to the experimental group, and only different amounts of boric acid were added. The relationship between the amount of added boric acid and the yield of crude magnesium is shown in Fig. 4.22. The

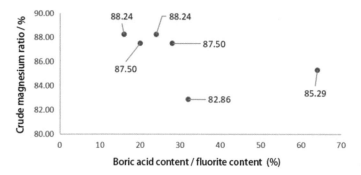

Fig. 4.22 Influences that using H_3BO_3 to replace fluorite has on the yield of crude Mg

abscissa in the figure is the percent of boric acid that was added to the experimental group with respect to the amount of fluorite in the reference group.

Crude magnesium ratio = yield of crude magnesium in experimental group/yield of that in reference group

As seen in Fig. 4.22, completely replacing fluorite with boric acid failed to achieve the yield of fluorite mineralizer. When the amount of added boric acid was 40–70 g (the amount of added boric acid/that of fluorite added in the reference group is 16–28%), the yield of crude magnesium fluctuates around 87–88%. With a further increase in the amount of boric acid, the yield of crude magnesium decreased inversely. When the amount of added boric acid reached the maximum value of 160 g (boric acid/fluorite is 64%), the crude magnesium yield was only about 85%. For the reference group, the crude magnesium output that obtained via magnesium smelting with fluorite was higher than that of the group in which boric acid was used as a mineralizer. To prevent boric acid from affecting the quality of crude magnesium that was produced via magnesium smelting, an X-ray diffractometer was used to perform elemental analysis on all crude magnesium products that were obtained in the experiments, and a high precision fluorimeter (fluoride ion selective electrode) was used to detect the fluorine content of samples. The test results confirm that the quality of the experimental crude magnesium samples was good and meets the standards for magnesium smelting products.

In experiment 1 (b), the two reduction retorts were loaded with same amounts of dolomite and ferrosilicon. To facilitate the experiment, the amounts of raw materials were reduced and the proportions of the raw materials remained unchanged. The sample in the reference group still followed the production formula and used fluorite as a mineralizer. The difference from experiment 1 (a) is that in experimental 1 (b), both fluorite and boric acid were added to reduce the amount of fluorite, which reduces fluorine pollution. The experimental data is shown in Table 4.7.

As clearly seen from the data in Table 4.7, the magnesium smelting experiment that used boric acid to partly replace fluorite achieved remarkable results. Under the same conditions, the crude magnesium outputs of most experimental groups were equal to or greater than that of the reference group that used fluorite as a mineralizer.

Table 4.7 Experimental data of magnesium smelting with boric acid used to replace some of the fluorite

Experimental series		Proportion of mineral in raw materials/%			Oversize rate of Mg slag/%	Crude Mg ratio/%
		Fluorite	Boric acid	Total		
B1	Reference group	2.1	0	2.1	44	
	Experimental group	0.85	0.42	1.27	76	100
B2	Reference group	2.13	0	2.13	60	
	Experimental group	0.86	0.69	1.55	92	87.5
B3	Reference group	2.1	0	2.1	80	
	Experimental group	0.43	0.39	0.82	92	100
B4	Reference group	2.4	0	2.4	60	
	Experimental group	0.49	0.24	0.73	76	108
B5	Reference group	2.49	0	2.49	28	
	Experimental group	0.51	0.25	0.76	88	116.7
B6	Reference group	2.46	0	2.46	52	
	Experimental group	0.5	0.21	0.71	89	100
B7	Reference group	2.46	0	2.46	60	
	Experimental group	0.5	0.50	1.00	84	96.0
B8	Reference group	2.46	0	2.46	52	
	Experimental group	0.5	0.50	1.00	80	92.3
B9	Reference group	2.46	0	2.46	52	
	Experimental group	0.5	0.60	1.10	89	100

Except for groups B1 and B2, the amount of added fluorite in the experimental group was only about 20% of that in the reference group. With added boric acid, the total amount of mineralizer was still less than the amount of fluorite mineralizer in the reference group. The reduced amount of fluorite in the mineralizer directly reduces the fluorine emission and relieves stress on the environmental pollution.

The oversize rate of magnesium slag refers to the ratio between the weight of slag that remained on the sieve and the total weight of magnesium slag after the magnesium slag was passed through a 0.45 mm sieve (the weight of remaining Mg slag/total weight of Mg slag). The oversize rate is the numerical data used to measure the degree of pulverization of magnesium slag. In Table 4.8, the oversize rates of all of the reference groups are lower than those of the corresponding experimental groups, and this indicates that the magnesium slag in the experimental group has better volume stability. Figure 4.23 shows photos of magnesium slag in the reference group and in the boric acid-doping experimental groups after they were stored for a period of time. In panel (a), the magnesium slag was completely powdered, and in panel (b), the magnesium slag remained lumpy.

Experiment II: using rare earth oxides to replace fluorite. The raw materials used in Experiment II were exactly the same as those used in Experiment I. Different rare earth oxides (cerium oxide (CeO_2), lanthanum oxide (La_2O_3), neodymium oxide (Nd_2O_3), yttrium oxide (Y_2O_3), and terbium oxide (Tb_2O_3)) were used as mineralizers to replace fluorite. The experimental process was also the same as before. The

Table 4.8 Formula for the industrial fluorine-free magnesium smelting test

Number of test 编号	Dolomite (kg)	Ferrosilicon (kg)	Fluorite (kg)	Type of mineralizer	Amount of mineralizer (%)
1	100	20.5	0	Boric acid	0.8
2	100	20.5	1.2	Boric acid	1.8
3	100	20.5	0	CeO_2 waste	0.8
4	100	20.5	1.2	CeO_2 waste	1.8

Fig. 4.23 Morphology of Mg slag after the smelting experiment: **a** reference group using fluorite **b** experimental group using boric acid

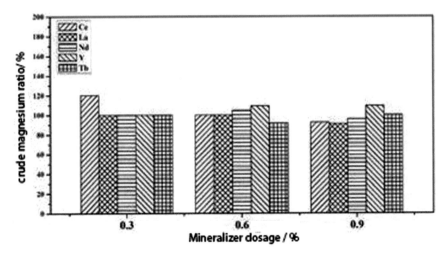

Fig. 4.24 Crude Mg yield of magnesium smelting using rare earth oxides as mineralizers

results indicate that when a rare earth oxide was used as a mineralizer in magnesium smelting, both the crude magnesium yield and the quality of crude magnesium were not inferior to that obtained using a fluorite mineralizer; also, the volume stability of magnesium slag was enhanced, and no powdering occurred. The crude magnesium ratios in the magnesium smelting experiments that used several rare earth oxides are shown in Fig. 4.24. In the figure, Ce, La, Nd, Tb, and Y denote CeO_2, La_2O_3, Nd_2O_3, Y_2O_3, and Tb_2O_3, respectively. The amount of mineralizer on the abscissa refers to the weight percent of the added rare earth mineralizer in the raw materials.

As seen in Fig. 4.24, when the amount of added rare earth mineralizer was 0.3% of the total weight of the raw materials, the yields of crude magnesium in the experiment samples with rare earth oxides were 100% of that of the sample with fluorite mineralizer, except for CeO_2. CeO_2 is a particularly prominent mineralizer used in magnesium smelting, and results in a crude magnesium ratio of 120%. Meanwhile, the amount of fluorite in the reference group was 2.5% of the total amount of raw materials, and this is times higher than that of a rare earth mineralizer. When the amount of added mineralizer was 0.6%, CeO_2 and Y_2O_3 showed greater advantages, and the corresponding crude magnesium yields were 105% and 109%, respectively. The crude magnesium rates of the remaining samples that used rare earth oxides also reached 100%, except for Tb_2O_3 which was only 92%. When the amount of added mineralizer was 0.9%, Y_2O_3 still maintained better performance with a crude magnesium yield of 109% whereas those of other rare earth oxides were not as good as that of the reference group, which used fluorite mineralizer. In the magnesium smelting experiment using rare earth oxides to replace fluorite as a mineralizer, the oversize rate of magnesium slag was consistent with the result of Experiment I; that is, when rare earth oxides were used as mineralizers, magnesium slag showed good volume stability. However, for the reference sample that used fluorite as a mineralizer, the magnesium slag was instantly efflorescence when it was removed from the furnace.

Although a variety of rare earth oxides as mineralizers exhibit good effects in magnesium smelting, it is not practical to use them in large-scale production because of their high prices. However, the recycled waste of a CeO_2 polishing agent, which is used as an abrasive for polishing optical glass, still contains many rare earth oxides. From the perspective of reducing the costs of raw materials and recycling industrial by-products, using recycled CeO_2 polishing agent as a mineralizer for magnesium smelting instead of using fluorite has attractive prospects. The following section introduces in detail a magnesium smelting experiment using recycled CeO_2 polishing agent instead of fluorite.

Magnesium smelting experiment using recycled CeO_2 instead of fluorite. In this experiment, the raw materials for magnesium smelting were the same as those in previous experiments. The mineralizer of the experimental group was recycled CeO_2 polishing agent (provided by Ganzhou Jinchengyuan New Material Co., Ltd.), and the mineralizer of the reference group was fluorite. The recycled cerium oxide polishing agent is denoted C2. Phase analysis shows that the composition of C2 is 32.58% rare earth oxides, 31% Al_2O_3, 18% SiO_2, 5.73% MgO, 4.57% F, and 3.97% CaO. XRD data and particle size distribution results are shown in Fig. 4.25; panel (a) shows the phase analysis XRD data of C2, and panel (b) shows the particle size distribution of C2.

The experimental result for the case when C2 was used to replace fluorite is shown in Fig. 4.26. The abscissa is the ratio between the amount of added C2 in the experimental group and that of added fluorite in the reference group. The ordinate is

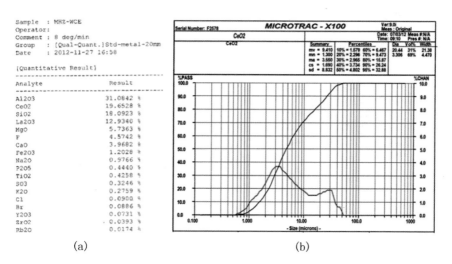

(a)   (b)

Fig. 4.25 Phase and particle size characterization of C2: **a** XRD data and **b** size distribution

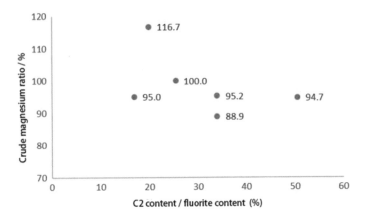

Fig. 4.26 Magnesium smelting using C2

the ratio between the crude magnesium yield of the experimental group and that of the reference group.

Figure 4.26 shows the results of the magnesium smelting experiment using C2 instead of fluorite. Compared with the results in Fig. 4.22 for the experiments that used boric acid to replace fluorite, it is clear that C2 is better than boric acid for use as a mineralizer in magnesium smelting. There are two sets of data that can meet or exceed the mineralizing effect of fluorite, and the amount of mineralizer was only 20–25% of the amount of fluorite. However, in industrial applications, a stable supply source of C2 needs to be further investigated. The morphology of the crude magnesium when recycled cerium oxide polishing agent was used as a mineralizer in magnesium smelting is shown in Fig. 4.27.

Figure 4.28 shows the XRD patterns of magnesium smelting slags with different amounts of added C2. The fluorite percent of the sample in the reference group

Fig. 4.27 Morphology of crude magnesium from pilot experiment **a** reference group using fluorite **b** experimental group using C2

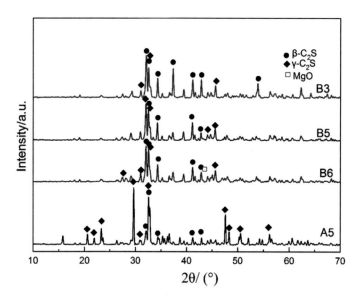

Fig. 4.28 Comparison of XRD patterns for magnesium slag using C2

A was 2 wt%, and that of C2 in experimental groups B3, B5, and B6 were 1.01, 0.68, and 0.51%, respectively. As seen from the comparison between the XRD phase analysis of the reference group A5 and the experimental groups B6, B5, and B3, the characteristic peaks of γ-C_2S almost disappeared, and the characteristic peaks of β-C_2S were significantly enhanced in the experimental groups. The changes show that the content of γ-C_2S decreased, whereas the content of β-C_2S correspondingly increased in magnesium slag when recycled cerium oxide was used as a mineralizer. The added amount of recycled cerium oxide in the experimental group B3 was the highest, at a value of 1.01%, which is still about half the amount of fluorite in the reference group. Correspondingly, the content of β-C_2S in the magnesium slag of B3 was the highest, and this magnesium slag sample had the best stability.

In summary, this pilot magnesium smelting test confirmed that the output and quality of crude magnesium that were obtained were basically consistent whether recycled cerium oxide or fluorite was used as a mineralizer. When the amount of added recycled cerium oxide was varied from 0.5 wt% to 1.0 wt% of the raw materials, there was little effect on the output of crude magnesium, but there was a significant impact on the stability of the magnesium slag. The magnesium slag was fragile when the added amount was less than 0.5 wt%, whereas the stability of magnesium slag became good when the added amount was above 0.7 wt%. As seen from the XRD pattern, using C2 as a mineralizer can prevent the crystal phase transition from β-C_2S to γ-C_2S, which stabilizes the magnesium slag. From a comprehensive consideration of the output and the quality of crude magnesium, and stability of magnesium slag, it was found that the optimal effect was achieved when the amount of added C2 was 0.7-1 wt% of the total raw materials. Using C2 instead of CaF_2 in magnesium smelting lays a good foundation for the industrial production of new clean magnesium smelting technologies and also achieves the reuse of rare earth waste; thus, this approach has promising application prospects in resource conservation and environmental protection.

4.2.6   Industrialized Test of Fluorine-Free Mineralizer

On the basis of the pilot test, boric acid (industrial grade) and recycled cerium oxide polishing agent were selected as fluorine-free mineralizers to replace fluorite. Industrial tests were conducted at Ningxia Huiye Magnesium Group Co., Ltd. The industrial test was implemented in accordance with the standard Pidgeon process, which has a technological flow as follows: dolomite calcination, batching, powder making, pelleting, packaging (pellets in paper bags), smelting conducted in reduction furnace, crude magnesium collected and refined, ingot costing, quality analysis, and final products are packaged.

Experimental formula: The formulas for the industrial fluorine-free magnesium smelting test are shown in Table 4.8. The amounts of mineralizer that are shown in the table are the percent of mineralizer in the total raw materials

Experimental process: The batching process combined manual and automatic batching systems. The pelleting process was completed using a pelletizer at Huiye's Magnesium No. 1 Branch Plant No.1 workshop. **Reduction smelting** was carried out using the production reduction furnace at Huiye's Magnesium No. 1 Branch Plant. The main working conditions during the test include: reduction temperature, degree of vacuum, water temperature, and sintering time; these factors were all implemented in accordance with the present production process standards of Huiye Magnesium Group Co., Ltd. **Measurement of crude magnesium output:** The input-output rates of every experimental batch were measured, recorded, and calculated separately. **Quality analysis of refined magnesium ingots:** The experimental crude magnesium was transported to the refining workshop for refining, separated and the quality and composition of the product were analyzed. **Monitoring and analysis of reduction slag:** The experimental batch of magnesium slag was sampled separately to determine the oversize rate and the contents of various components in the slag.

Experimental results: The output and yield of a single retort in the test batch were basically the same as those of the normal production data. Through quality analysis of the magnesium ingot, it was found that the experimental ingot meets all of the quality specifications for the product and that the morphology of the magnesium slag is basically all granular without powdering. Compared with the slag that used fluorite as a mineralizer, the slag in this experiment greatly reduced dust pollution to the environment. Figure 4.29 shows the magnesium slag generated in the industrial test, where panel (a) shows the magnesium slag that used boric acid to replace fluorite and panel (b) shows the magnesium slag that used C2 to replace fluorite.

Fig. 4.29 Morphology of magnesium slag after the industrial test: **a** using boric acid as a mineralizer and **b** using C2 as a mineralizer

Conclusions:

The industrial test of fluorine-free magnesium smelting confirms that boric acid and C2 as fluorine-free mineralizers can be used to replace fluorite, which is usually used in magnesium smelting. This substitution can be made without affecting the output and production rate of magnesium products. The replacement does not change the original production process and does not increase the production costs. Furthermore, the stability of the slag is greatly enhanced, and so the problems of dust pollution during the collection, transport, and reuse of slag have been solved. More importantly, this approach reduces the use of fluoride from the natural sources and reduces fluoride pollution in soil, water, and air.

4.2.7 Application of Fluorine-Free Mg Slag in Steel-Making

Using the fluorine-free magnesium slag to partially replace limestone as a slagging agent in steel-making reduces production costs, enables the recycling and use of industrial waste, and reduces the stress that magnesium slag has on the environment. Fluorine-free magnesium slag was first used to conduct a small laboratory test to confirm that there is no adverse effect on the quality of the simulated steel-making product, and then an industrial test was carried out in the steel plant.

Laboratory simulation experiments were carried out at the China Central Iron and Steel Research Institute [20]. Fluorine-free magnesium slag from the Heiye magnesium Ltd. was used. This slag basically does not contain sulfur and phosphorus. The laboratory experiment uses low-S low-P iron, fluorine-free magnesium slag, and converter steel slag from the steelmaking production of Ningxia Steel as the raw materials. A certain ratio of magnesium slag and converter steel slag were mixed with iron powder, heated, melted, and held at a high temperature for a while to simulate converter steelmaking process and to investigate the change of sulfur and phosphorus content in iron-containing materials. **Experimental formula:** The experimental materials included iron material and 10 wt% slags(steel slag + magnesium slag). 5, 10, 15, and 20% of the total slag of fluorine-free magnesium slag were used separately with steel slag to study the effect that magnesium slag has on steel making. **Experimental process:** The mixed powder was kept at 1550 °C for 40 min. After the experiment, molten steel and slag were discharged from a crucible and cooled in the air. The contents of C, Si, Mn, P, and S in the iron material were analyzed. **Experimental results:** With an increase in amount of added magnesium slag, the S content in the steel sample decreased, and the P contents only increased slightly. When the magnesium slag was 15 and 20%, the P contents of the samples were basically the same as that of the original iron powder. The results indicate that magnesium slag has no adverse effects on the quality of molten steel, especially the S and P in molten steel when magnesium slag is used to replace some of the steel slag as the slagging agent (The replacement ratio of magnesium slag to steel slag was controlled to be within 20%.). Since magnesium slag has lower melting point, partial replacement of the slagging agent by magnesium slag is beneficial to converter slagging and increases the melting rate of the steel-making slagging agent. Because of laboratory conditions, the actual production effect on site will be better than the laboratory results.

Industrial test A further production test was carried out under actual production conditions at the Ningxia Steel Plant. Test scheme: A separate hopper for magnesium slag was used. Magnesium slag was added simultaneously when the first batch of slagging agent was added using the feeding platform. The amounts of added magnesium slag are 5, 10, 15, and 20%. In the first batch, there were three furnace tests: (1) Slagging agent was added according to normal steel-making production, and an additional 10% magnesium slag was added simultaneously when the first batch of slagging agent was added. (2) The amount of steel-making slagging agent was decreased by 10%, and 10% magnesium slag was added instead of the slagging agent. (3) The steel-making slagging agent was reduced by 20%, and 20% magnesium slag was used instead of the slagging agent. The experimental results indicate that without changing the present process conditions, magnesium slag can be used to partially replace lime as a slagging agent in steel making and that doing so has no negative impact on the quality of the finished steel products until the amount of substituted magnesium slag reaches 15%. In the second batch, the maximum percent of substituted magnesium slag was as high as 30%, and the effect was the same.

References

1. Kim YJ, Nettleship I, Kriven W et al (1992) Phase transformations in dicalcium silicate: II, TEM studies of crystallography, microstructure, and mechanisms. J Amer Ceram Soc 75(9):2407–2419
2. Akira S, Yoshio A et al (1986) Development of dusting prevention stabilizer of stainless steel slag. Kawasaki Steel Giho 18:20–24
3. Chan C, Kriven WM et al (1992) Physical stabilization of the β → γ transformation in dicalcium silicate. J Am Ceram Soc 75(6):1621–1627

4. Yang Q et al (2008) Stabilization of EAF slag for use as construction material. In: REWAS 2008: global symposium on recycling, waste treatment. Minerals, Metals and Materials Society, pp 49–54
5. Tossavainen M, Engström F et al (2007) Characteristics of steel slag under different cooling conditions. Waste Manage 27(10):1335–1344
6. Jiang H, Li L et al (2012) Study on aging phase transformation and property of air-quenching steel slag. Bullet Chinese Ceramic Soc 31(01):171–174 + 192
7. Zhu G (2014) Research on methods to suppress the expansion of steel slags and road performance used as roadbed filling material. Dissertation, Nanjing Forestry University
8. Cui Z, Ni X et al (2006) Study on the expansibility of magnesium slag. Fly ash Compreh Utilization 06:8–11
9. Jürgen G (2000) Properties of iron and steel slags regarding their use. In: Paper presented at the 6th international conference on molten slags, fluxes and salts, Stockholm City, Stockholm, Sweden-Helsinki, Finland, 12–17 June
10. Du C, Wu L et al(2012). Dust pollutionpreventation of magnesium slag. In: Paper presented at the 8th conference of Chinese society of particuology, Sept 5–8, Hangzhou, China
11. Wu L, Yang Q, Han F et al (2013) Dust control of magnesium production by Pidgeon process. In: Paper presented at the 7th international conference on micromechanics of granular media, Sydney, 8–12 July
12. Feng J, Long S et al (1985) Effect of minor ions on the stability of β-C_2S and its mechanism. J Chinese Ceramic Soc 04:424–432
13. Zhang W, Zhang J et al (2019) Structure and activity of dicalcium silicate. J Chinese Ceramic Soc 47(11):1663–1669
14. Huang W, Wen Z et al (2018) Effects of phosphorus and sulfur doping on the crystal structure of dicalcium silicate. Bulletin of the Chinese ceramic society 37(08):2502–2505 + 2511
15. Han F, Yang Q et al (2013) Reclaim and treatment of magnesium slag from pidgeon process. Inorgan Chem Indus 45(7):52–55
16. Han F, Yang Q, Wu L et al (2011) Treatments of magnesium slag to recycle waste from pidgeon process. Adv Mater Res 418–420:1657–1667
17. Barua S, Wynnyckyj JR et al (1981) Kinetic of the silicothermic reduction of calcined dolomite in flowing hydrogen. Can Metall Q 20(3):295–306
18. Feng G et al (2009) Life cycle assessment of primary magnesium production using the pidgeon process in China. Int J Life Cycle Assess 14:480–489
19. Wu L, Han F, Hang Q et al (2012) Fluoride emissions from Pidgeon process for magnesium production. In: Paper presented at the 27th international conference on solid waste technology and management, Philadelphia, PA U.S.A, March 11–14
20. Han F etc (2012) National international science and technology cooperation. Study on the comprehensive treatment and recycling technology of magnesium slag(2010DFB50140)Project Technical Report

Chapter 5
Resource Utilization of Magnesium Slag

Abstract The industrial solid waste recycling team at North Minzu University has done much research on resource utilization of magnesium slag. This chapter mainly introduces a discussion of resource utilization of magnesium slag. This includes the use of magnesium slag to prepare glass ceramics, porous ceramics, and sulfoaluminate cement clinker, fixing/stabilizing heavy metals in acidic residue generated by lead-zinc smelting, and modifying copper slag.

Keywords Slag based glass ceramics · Porous ceramics · Sulfoaluminate cement clinker · Heavy metals

5.1 Preparation of Glass Ceramics Using Mg Slag

5.1.1 Preparation Principle of Glass Ceramics

Glass ceramics are uniform polycrystalline materials in which glass and crystals coexist. They are commonly prepared via the addition of a crystal nucleating agent, forming crystal nuclei in the glass after proper heat treatment followed by crystal nucleus growth; ultimately, glass ceramics are obtained. Compared with other materials, glass ceramics have properties such as an adjustable thermal expansion coefficient (a zero expansion coefficient can be achieved), high mechanical strength, excellent electrical insulation, low dielectric loss, remarkable wear resistance, corrosion resistance, high temperature resistance, and also good chemical stability [1].

Using magnesium slag to prepare glass ceramics can consume and reuse a large amount of industrial waste that is generated by magnesium smelting. Furthermore, the prepared glass ceramics materials have excellent properties. Han Fenglan of North Minzu University prepared glass ceramics via sintering. Magnesium slag, resin ash, and alumina were mixed according to the composition design, and the mixture was melted at high temperature, quenched with water, and ground to fine powder. After the thermal performance was characterized using thermogravimetric analysis and differential scanning calorimetry (TG-DSC), the glass powder was dried, pressed into shape, and then the sample was nucleated (hold time of 1 h) and crystallized (hold times of 0.5, 1, 1.5, 2, and 2.5 h) to prepare glass ceramics [2].

© The Author(s) 2021
L. Wu et al., *Comprehensive Utilization of Magnesium Slag by Pidgeon Process*, SpringerBriefs in Materials, https://doi.org/10.1007/978-981-16-2171-0_5

5.1.2 Development of Glass Ceramics in China and the World

Glass ceramics were originally developed from photosensitive glass in 1957. Since Kemantaski in the United Kingdom first used blast-furnace slag to prepare glass ceramics in 1965 [3], glass ceramics were rapidly developed from specific application to extensive uses in architecture and other fields [4]. With the development of the technological economy, glass ceramics have been successfully commercialized. In Europe and America, rock glass ceramics and slag glass ceramics were first industrialized for use in decoration materials in buildings. In the mid-1960s, the former Soviet Union reported that slag glass ceramics could be used as building materials in practical production. In the early 1970s, Czechoslovakia used molten-cast basalt to manufacture wear-resistant flooring materials. At the same time, the production of rock-ceramic glass for use as decorative panels in construction also emerged in United States [5, 6]. In Western countries, lithium-based glass ceramic materials have been used to manufacture optical fiber connectors. The thermal expansion coefficient and hardness of lithium-based glass ceramics match those of optical fiber connecters made of silica glass better than traditional zirconia. In addition, lithium-based glass ceramics are more easily machined with high precision and have excellent environmental stability.

In Asia, Japan was the first to develop and put glass ceramics into practical use for buildings; the melt-sintering method was mainly used to produce artificial marbles with a glass ceramic structure. South Korea also produces high-grade glass ceramic decorative panels. In China, glass ceramics are well-developed and are used in building materials and mechanical engineering because, in addition to other advantages, they have good corrosion resistances, wear resistances, good insulation, light specific gravity, and can be seamlessly welded with metals. The steel slag-based glass ceramics that have been developed via sintering formation in China are compact and have a smooth surface and low porosity; thus, they have promising market values. Slag glass ceramics "consume waste slag" and also have the advantages of excellent performance, a simple preparative technique, and practical scalability production; hence, it has always attracted the attentions of material scientists and engineers in various countries. In recent years, the use of slags to prepare glass ceramics as a reasonable method to recycle slag has attracted increasing attention [6–10].

5.1.3 Preparation of Glass Ceramics Using Slags

The preparation process of glass ceramics generally includes melting, sintering, and the sol-gel method [11, 12]. Among these, melting and sintering are the main preparation methods.

Melting method

Melting is the first method proposed for preparing glass ceramics. The main process is to mix and grind the nucleating agent with other raw materials first, and then the mixture is melted in a high-temperature environment with a temperature range from 1300 to 1600 °C. After the mixture is completely melted, homogenized, and molded, it is annealed and then nucleated and crystallized using a certain heat treatment process to obtain a uniform and dense glass ceramic sample.

Sintering method

The sintering method was first proposed by Xuanboen around 1960 to produce glass ceramics; this method was used to successfully implement industrial production in Japan around 1970. The process flow is as follows: raw materials are mixed, melted, water quenched, smashed, sieved, shaped, sintered, and deep processed, and thus, products are obtained.

The sintering method overcomes the drawbacks of the melting method, which involves inseparable melting and molding, difficulties in controlling molding process at high temperature, and the need for a nucleating agent. Thus, the prepared glass ceramics have better plasticity. Therefore, the ceramics are more suitable for molding at low temperatures and favorable for industrial production. In addition, the water-quenched glass ceramics exhibit a better overall crystallization phenomenon [13, 14].

5.1.4 Experimental Preparation of Glass Ceramics Using Mg Slag

In this experiment, magnesium slag, resin ash, and alumina were used to prepare glass ceramics via sintering. Magnesium slag was provided by Ningxia Huiye Magnesium Co., Ltd., resin ash powder was provided by a foundry in Ningxia, and alumina powder was provided by Tianjin Guangfu Fine Chemical Research Institute. The compositions of the raw materials are shown in Table 5.1. According to the compositions of the raw materials, the glass ceramic system selected for the experiment is the CMAS (CaO-MgO-Al_2O_3-SiO_2) system; the contents of magnesium slag, resin ash, and alumina are 48.78, 48.78, and 2.44%, respectively [2].

The raw materials were weighed in proportion, placed in a mortar, and manually ground for about 20 min to prepare the mixture. The prepared mixture was placed in a

Table 5.1 Compositions of main raw materials

Raw material	Composition/%					
	SiO_2	CaO	MgO	Al^{3+}	F^-	Mn^{2+}
Mg slag	28.56	50.65	3.84	0.73	7.74	2.44
Resin ash	90.18	1.56	0.01	3.69	0.01	0.01

corundum crucible suspended in the middle of a vertical tube furnace; the temperature was increased to 800 °C at a heating rate of 8 °C/min, and the temperature was maintained at 800 °C for 2 min. Next, the temperature was increased to 1385 °C at a heating rate of 5 °C/min, and then, that temperature was held for 90 min.

At the end of the temperature holding time, the corundum crucible and the molten glass were quenched in a water tank located below the vertical tube furnace. This allows the high-temperature molten glass to be instantly cooled. After cooling, the glass sample and the corundum crucible were placed in a drying oven at 100 °C for 20 min. The prepared glass was removed from the corundum crucible and ground in a vibration mill for 30 s to obtain glass powder that had a size of about 400 mesh (37 μm).

To determine the parameters of following sintering process, the obtained glass powder was analyzed using TG-DSC. In sintering process, the powder was pressed into discs that had a diameter of 25 mm under 100 MPa in a dry press for 40 s. The discs were then placed in a vertical tube furnace for sintering. The temperature was increased from room temperature to 500 °C at a heating rate of 10 °C/min. After this temperature was held for 5 min, the temperature was increased to 845 °C at a heating rate of 5 °C/min; this temperature was held for 60 min, and then it was increased to 1021 °C at a heating rate of 3 °C/min. After 150 min of temperature holding, the samples were naturally cooled, and glass ceramics that were made from magnesium slag were obtained.

5.1.5 Property, Morphology, and Phases of Mg slag based Glass Ceramics

The TG-DSC diagram of the prepared glass powder is shown in Fig. 5.1, and the detection instrument was a German Netzsch STA449 thermal analyzer. Glass

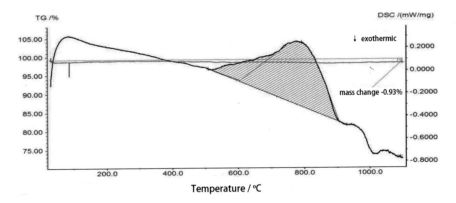

Fig. 5.1 TG-DSC diagrams of glass powder samples

ceramics need to absorb relatively high energy during nucleation, and so, the temperature 775 °C that corresponds to the maximum endothermic peak is determined to be the nucleation temperature of the glass ceramics. The internal state of glass ceramics tends to be stable during crystallization; the internal energy is subsequently released to form crystals. Therefore, the energy of the glass ceramics is lowest when the ceramics are in the crystalline state. As seen, the trough corresponding temperature in the DSC spectrum is 1021 °C, and this is the crystallization temperature.

Volume density ρ (g/cm^3) can be calculated according to following equation:

$$\rho = m_0 w/(m_1 - m_2)$$

where:m_0 denotes the mass of dry sample in air (g); m_1 denotes the mass of water-saturated sample in air (g); m_2 denotes the mass of water-saturated sample in water (g); w denotes the density of distilled water at room temperature (g/cm^3)

Water adsorption rate Wa (%) is calculated according to the following equation:

$$Wa = (m_1 - m_0)/m_0 \times 100\%$$

The changes in the volume density of the glass ceramic sample with respect to the holding time are shown in Fig. 5.2. As seen in Fig. 5.2, when the holding time was longer, the volume density gradually increased. In the interval from 0.5 to 1 h, the volume density of the sample basically remained constant. From 1 h to 2 h, the volume density of the sample increased rapidly until it reached a maximum value of 2.77 g/cm^3 when the holding time was 2 h. Conversely, the volume density of the sample decreased slightly util the holding time reached 2.5 h.

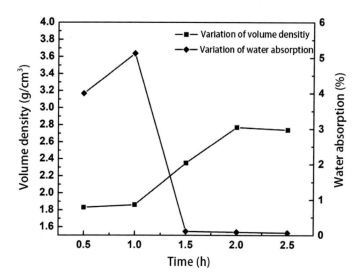

Fig. 5.2 Relationships between volume density, water adsorption rate, and temperature holding time

In the period from 0.5 to 1.5 h, the water adsorption rate of the sample first increased and then decreased. After 1.5 h, the decreasing trend slowed down, and the water adsorption rate was almost parallel to the X axis. The data shows that with holding time longer than 1.5 h, the internal structure of the glass ceramic sample became denser, and the water absorption rate was very low. The water absorption rate reached a maximum value of 5.13% when the holding time was 1 h. When the holding time was 2.5 h, the water adsorption rate of the sample reached the lowest value of only 0.07%.

The samples with holding times of 1, 1.5, 2, and 2.5 h were scanned using a scanning electron microscope (Zeiss EVO-18, Germany), and the microstructures were observed. SEM images of the glass ceramics with different crystallization holding times are shown in Fig. 5.3. In the figure, panels (a), (b), (c), and (d) show the sample microstructures when the holding times were 1, 1.5, 2, and 2.5 h, respectively.

It is clear from Fig. 5.3 that with longer crystallization holding time, the glass ceramics made from magnesium slag became increasingly denser, and the porosity decreased. It can be seen that the extended crystallization holding time enhanced the density of glass ceramics.

In Fig. 5.4, the upper left pattern corresponds to the sample that had a holding time of 0.5 h; the upper right pattern corresponds to the sample that had a holding time of 1.5 h; the bottom pattern corresponds to the sample with a holding time of

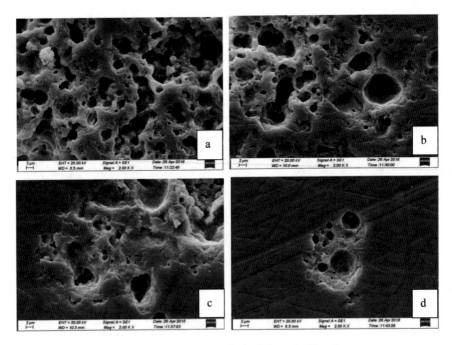

Fig. 5.3 Microstructures of Mg slag glass ceramics in different holding time

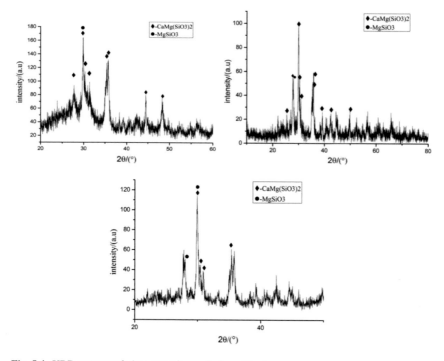

Fig. 5.4 XRD patterns of glass ceramics made from Mg slag

2.5 h. It can be seen that the main crystal phase in Mg slag glass ceramics is diopside $(CaMg (SiO_3)_2)$ with a small amount of inclined top flint $(MgSiO_3)$.

5.2 Preparation of Porous Ceramics Using Mg Slag

Porous ceramics are a new type of ceramic material. Porous ceramics are sintered at a high temperature and have a large number of interconnected or closed pores that form in the material body during the formation and sintering process. The advantages of porous ceramic materials are that they are environmentally friendly and have high porosity, low air resistance, stable chemical properties, good regeneration performance, and good resistances to high temperature, high pressure, and chemical corrosion. Thus, porous ceramic materials have been rapidly developed in recent years and have been widely used in filtration, purification and separation, catalyst carriers, sound and shock absorption, thermal insulation materials, biological materials, sensors, and aeronautical and aerospace materials [15–17].

5.2.1 Characteristics of Porous Ceramics

(1) High porosity. Porous ceramics are characterized by a large number of uniform pores with controllable sizes. The pores are classified as open pores and closed pores. The functions of open pores include filtering, absorbing, adsorbing, and eliminating echoes, while closed pores are favorable for insulating heat and sound, and transfer of liquid and solid particles.

(2) High strength. Porous ceramics are generally prepared from metal oxides, such as silicon dioxide and silicon carbide, and are sintered at high temperatures. All of these materials have intrinsic high strength. During the calcination process, the boundary regions of these raw material particles melt and bond, thereby forming ceramics that have higher strength.

(3) Stable physical and chemical properties. Similar to other ceramic materials, the advantages of porous ceramics are that they have good acidic and alkaline corrosion resistance and withstand high temperature and high pressure. In addition, porous ceramics have a good self-cleaning ability and do not cause secondary pollution. Thus, they are green and environmentally friendly functional materials.

(4) High filtration accuracy and good regeneration performance. Porous ceramic materials that are used as filter materials have narrower pore size distribution, higher porosity, and higher specific surface area. The porous ceramics fully contact the filtered materials, and pollutants such as suspended solids, colloids, and microorganisms are blocked on the surface or inside of the filter medium, exhibiting a better filtering effect. After a period of use, the porous ceramic filter material can be backwashed with gas or liquid to restore its original filter capacity.

5.2.2 Preparation of Porous Ceramics Using Mg Slag

5.2.2.1 Selection of Raw Materials

Porous ceramics that are commercially available mostly use materials such as Al_2O_3, SiC, and mullite, as the main raw materials. These materials have relatively high prices and complicated preparation processes, and these characteristics limit the promotion and applications of porous ceramics [16, 17]. The main components of magnesium slag, fly ash, and acetylene sludge are SiO_2, CaO, and Al_2O_3, which are similar to the components of commercially available porous ceramics materials; thus, solid wastes like magnesium slag can be used as raw materials to prepare porous ceramics [18–20]. Magnesium slag is mainly composed of CaO, SiO_2, and Al_2O_3, and fly ash is mainly composed of CaO, SiO_2, Al_2O_3, and MgO. Fly ash is the main solid waste discharged from coal-fired power plants and one of the biggest displacement industrial wastes in China. Acetylene sludge is solid waste obtained after acetylene is prepared from calcium carbide, and it is composed mainly of CaO

Table 5.2 Compositions of porous ceramics

Name of solid waste	Compositions/wt%						
	CaO	SiO$_2$	MgO	Fe$_2$O$_3$	Al$_2$O$_3$	C	Ca(OH)$_2$
Mg slag	45–50	28–30	10–12	6–8	3–5		
Fly ash	5–7	40–45	5–6	8–10	20–24	8–10	
Acetylene sludge		6–8		6–8	10–12		75–80

and Ca(OH)$_2$. It also contains SiO$_2$ and Al$_2$O$_3$ and a small amount of CaCO$_3$, Fe$_2$O$_3$, MgO, carbon slag, and other components. Wanxiu et al. [21] used three industrial solid wastes (magnesium slag, fly ash, and acetylene sludge) as raw materials; the raw materials were mixed in a certain ratio, pressed, and sintered to obtain porous ceramics containing silicate phases.

5.2.2.2 Preparation Process of Porous Ceramics Using Magnesium Slag

The industrial solid waste magnesium slag used in the experiment was provided by Ningxia Huiye Magnesium Co., Ltd. The fly ash was provided by Ningxia Shenhua methanol plant. The acetylene sludge was from Ningxia Dadi chemical company. The chemical compositions of magnesium slag, fly ash, and acetylene sludge are shown in Table 5.2.

5.2.2.3 Properties of Porous Ceramics Made from Magnesium Slag

The raw materials were weighed according to different formulations then ground and mixed in a vibrating mill with a mixing time of 60 s. The mixture was unidirectionally pressed under a pressure of 63.7–95.9 MPa, and the holding time was 50-60 s. Pressureless sintering were carried out at a sintering temperature of 1150 °C, a heating rate of 10 °C/min, and a holding time of 4 h. After the sintering was complete, the samples were furnace-cooled to room temperature; then, the samples were removed, and their physical and chemical properties were measured.

5.2.2.4 Properties of Magnesium Slag Porous Ceramics

The experimental results of the sample that was sintered for 4 h at 1150 °C show that the prepared porous ceramic has a maximum compressive strength of 98 MPa when the ratio of magnesium slag to fly ash to acetylene sludge was 70:25:1. The porous ceramic has the largest porosity of 57% when the ratio of magnesium slag to fly ash to acetylene sludge was 12:3:5. When the magnesium slag:fly ash:acetylene sludge ratio was 6:3:1, the porous ceramics have a complete framework and uniformly distributed micropores. Adding acetylene sludge and carbon powder as pore-forming agents

Fig. 5.5 Cross-section photo
of Mg slag porous ceramics

can homogenize the pore distribution, refine pore size, enhance the porosity, and gas filtration performance of porous ceramics. With the same amount of pore-forming agent, the carbon powder has a better pore-forming effect; specifically, the weight loss, water absorption, and porosity of the porous ceramics reached maximum values of 30, 38, and 53%, respectively, and the minimum volume density was 1.4 g/cm³. The main phases in porous ceramics are mainly calcium metasilicate, dicalcium silicate, or calcium magnesium silicate, which are the products of the high-temperature reaction between CaO and SiO_2, and small amount of aluminosilicate and ferrite.

The typical microstructure of porous ceramics that are made from magnesium slag is shown in Fig. 5.5. For the sample in the photo, the ratio of magnesium slag:fly ash:acetylene sludge was 80:5:15. The weight loss of the sample was 10.33%; the porosity was 38.86%; the water absorption was 21.52%; the bulk density was 1.78 g/cm³; the compressive strength was 30.10 MPa.

5.3 Preparation of Slag-Sulphoaluminate Cement Clinker Using Mg Slag

Magnesium slag has a metastable high-temperature structure, and it has active cations and a high hydration activity. Liguang et al. [22] mixed magnesium slag, mineral slag, and cement clinker to prepare magnesium slag-based cementitious materials. The effects that the magnesium slag content, cement clinker content, grinding process, and auxiliary activator have on the strengths (compressive and flexural strength) of cementitious materials that are made from magnesium slag were discussed. The hydration products of magnesium slag cementitious materials were analyzed. Fenglan et al. [23] used magnesium slag to prepare cementitious materials. From this research, they summarized the following: magnesium slag has hydration activity and can form calcium silicate cement after hydration, and its strong hygroscopicity contribute to the durability and strength of cement mortar. Because of these characteristics, different contents of ground magnesium slag power were added to cement

(42.5) clinker. After studying its water consumption, setting time, flexural strength, compressive strength, and stability, it was concluded that it is feasible to add 15-20% of magnesium slag without affecting the cement performance indexes.

Sulphoaluminate cement has become one of the most active research hotspots because it has the characteristics of a low sintering temperature, low emission, high early-stage strength, good impermeability, and resistances to frost and corrosion. Ordinary sulphoaluminate cement is mainly prepared from limestone, alumina, and gypsum as raw materials. In recent years, to increase the resource utilization rate of industrial solid waste, researchers have conducted a lot of research in the preparation of belite sulfoaluminate cement using industrial solid wastes such as fly ash, coal gangue, and phosphogypsum as raw materials [24–27]. Zhao et al. [28] mixed electrolytic manganese residue and magnesium slag to sinter sulphoaluminate cement clinker, and then, they carried out a series of experimental studies, which are described in detail below.

5.3.1 Raw Materials of Slag-Sulphoaluminate Cement

Experimental materials: Electrolytic manganese residue was provided by Ningxia Tian Yuan Manganese Co., Ltd., and magnesium slag was provided by Ningxia Huiye Magnesium Co., Ltd. $CaSO_4 \cdot 2H_2O$ was from Tianjin Komiou Chemical Reagent Co., Ltd.; Fe_2O_3, SiO_2, Al_2O_3, and CaO are commercial reagents that were all analytically pure. As seen in Fig. 5.6, the phases in the electrolytic manganese residue (Fig. 5.6 a) are mainly $CaSO_4 \cdot 2H_2O$ and SiO_2. The phases in magnesium slag (Fig. 5.6 b) are mainly C_2S, $MgO \cdot Fe_2O_3$, $CaO \cdot MgO \cdot SiO_2$, of which C_2S is the primary part. Dihydrate gypsum and C_2S are the main components of sulphoaluminate cement. After calcination, dihydrate gypsum decomposes to form SO_3, and this is beneficial

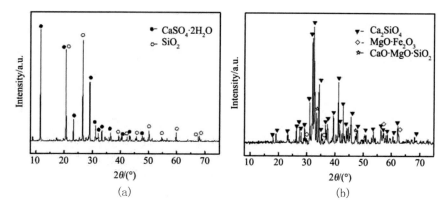

Fig. 5.6 XRD patterns of electrolytic manganese residue and magnesium slag: **a** phases in manganese slag and **b** phases in magnesium slag

Table 5.3 Composition design of slag-sulphoaluminate cement clinker

Alkalinity(Cm)	S/Al ratio(P)	Si/Al ratio(N)	β-C$_2$S	C$_4$A$_3$Š	C$_4$AF
1	4.23	1.85	40%	50%	10%

for the formation of calcium sulfoaluminate; C$_2$S can enhance the later strength of sulfoaluminate cement.

Sulfoaluminate cement clinker is composed of minerals such as anhydrous calcium sulfoaluminate (C$_4$A$_3$Š), dicalcium silicate (C$_2$S), and calcium aluminate ferrite (C$_4$AF). (In these abbreviations, the letters have the following designations: C-CaO, A-Al$_2$O$_3$, Š-SO$_3$, S-SiO$_2$, and F-Fe$_2$O$_3$). According to the mineral composition of sulfoaluminate cement and process control conditions, the designed composition of the experimental sample is shown in Table 5.3. The sintering temperatures are 1200, 1230, 1260, and 1300 °C, and the temperature was held for 30 min. To completely sinter the cement clinker in the muffle furnace, the sintering process was repeated twice.

5.3.2   Phases in Cement Clinker

X-ray diffraction was used to analyze the phase composition of the sintered sulphoaluminate cement clinker. Figure 5.7 shows the phases in the sulphoaluminate cement clinker sample that was sintered at different temperatures. The main mineral phases in the clinker samples that were sintered at 1200, 1230, and 1260 °C are C$_2$S and C$_4$A$_3$Š, and basically, there are no other phases. As seen from these three patterns, the intensities of the diffraction peaks of C$_2$S and C$_4$A$_3$Š in the sample that was

Fig. 5.7 XRD patterns of clinker sintered at different temperatures

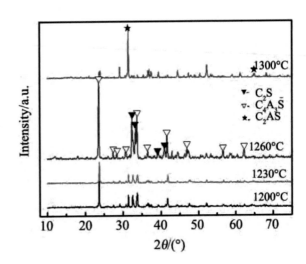

sintered at 1260 °C are much greater than those in the samples that were sintered at 1200 and 1230 °C. For the sample that was sintered at 1300 °C, the useless C_2AS with no hydration activity appears in the cement clinker, and its diffraction intensity is much higher than those of the useful components, such as C_2S and $C_4A_3\check{S}$. C_4AF is basically not observed in the figure because the iron phases in manganese slag and magnesium slag are equivalent to that in solid solutions (with lower melting point) during the calcination of cement clinker. Part of the solid solution enters into $C_4A_3\check{S}$, and this results in lower Fe content in C_4AF. Some of the intermediate product that forms C_4AF reacts with anhydrite to form $C_4A_3\check{S}$, and this leads to higher $C_4A_3\check{S}$ content in the clinker and a decrease in the C_4AF content [29, 30]. Therefore, through comprehensive analysis, it can be determined that 1260 °C is the optimal sintering temperature. Compared to the sintering temperature (1450 °C) of traditional Portland cement clinker, the sintering temperature in this experiment was much lower. The reason for this is that Fe oxides as solid solutions greatly reduce the sintering temperature of raw materials, and thus, slag that contains a small amount of iron phase reduces the sintering temperature of raw materials. Therefore, when manganese slag and magnesium slag are used to prepare sulfoaluminate cement clinker, the energy consumption can be reduced, the heating process can be easily controlled, and the sintering cost can be reduced.

5.3.3 Experimental Results

When different amounts of gypsum were added to the prepared cement clinker, a TAM Air eight-channel isothermal microcalorimeter (Retrac HB, Sweden) was used to determine the heat of hydration, to determine the optimal gypsum dosage, and to analyze the effect of different hydration times on the hydration performance. Compressive and flexural properties of the prepared sulphoaluminate cement clinker were tested according to the *Method of testing cements-Determination of strengths* (GB/T 17671-1999). The impermeability of the material was measured with reference to the impermeability test specification of cement mortar. According to the toxic leaching method in HJ/T 299-2007 *"Solid waste-Extraction process for leaching toxicity-Sulfuric acid and nitric acid method"*, sulfuric acid and nitric acid were used as the leaching agents to perform the toxicity leaching of different samples, and residues was determined. The concentration of heavy metal ions in the leaching solution was compared to the limit in the *Integrated Wastewater Discharge Standard* (GB 8978-1996).

The experimental results show the following: (1) The content of electrolytic manganese residue and magnesium slag in the raw meal can each reach 21%. The optimal sintering temperature was 1260 °C, and the holding time was 30 min. At this point, the mineral phase in the sintered sample was mainly C_2S and $C_4A_3\check{S}$. (2) A certain amount of gypsum was added to the prepared cement clinker. When the added amount was 15%, the hydration heat released reached a maximum, and the mechanical properties were the best. The flexural strength at 28 d was

5.1 MPa, and the compressive strength was 31.2 MPa. The impermeability level reached P6, and the sintered clinker and hydration products effectively solidified and stabilized the heavy metals in the industrial solid waste; this made it difficult for them to be leached. The following conclusions can be drawn: through sintering, the use of electrolytic manganese residue and magnesium slag can produce qualified fast-hardening sulphoaluminate cement clinker with higher early-stage strength. Compared to commercial silicate and sulphoaluminate cement clinkers, the cement clinker prepared using magnesium slag has advantageous properties, is low cost, can turning waste into treasure, and creates a low amount of pollution.

5.4 Solidification/Stabilization of Heavy Metals in Industrial Solid Waste

5.4.1 Fixation and Stabilization of Heavy Metals in Pb/Zn Smelting Slags

Solidification is a process in which appropriate additives are used in toxic and hazardous wastes to stably fix the toxic and hazardous substances in the waste. This eliminates pollution and damage caused by the waste to the surrounding ecosystem through reducing the leaching or release of toxic and hazardous components. Toxic and harmful pollutants are decomposed, precipitated, neutralized, or transformed into low-migration, low-dissolution, low-toxic, or even non-toxic matters through treatments; thus, environmental pollution caused by toxic and hazardous substances can be reduced. Because most of the materials used to treat toxic and hazardous solid wastes have both stabilizing and fixing effects, the decontamination and detoxication for waste is usually called stabilization/solidification technology, and this is sometimes shortened to solidification [31]. At present, stabilization/solidification technology has been widely used in the treatments for contaminated sites and solid waste. Compared with other technologies, such as chemical treatment or bio-remediation, stabilization/solidification technology has the advantages of convenient construction and low cost [32]. The curing agents that are used in stabilization/solidification technology include cement, slag, fly ash, quicklime, pharmaceuticals, organic polymers, geopolymers, modified clays, and certain waste materials [33]. The geopolymer is polymerized by inorganic silicon-oxygen tetrahedra and aluminum-oxygen tetrahedra. When the geopolymers are used to fix wastes, the process is simple and is more stable.

Chen et al. used magnesium slag and fly ash-based geopolymers to solidify/stabilize acidic residue generated via Pb-Zn smelting and conducted a large number of experimental studies [32, 34, 35]. These are described in detail below.

Lead and zinc smelting is an important case of China's nonferrous metal smelting. The bulk of industrial solid waste from lead and zinc industries in China mainly come from metallurgical slag and the residue that is generated via acidic water treatment (acidic residue). The amount of annual industrial waste that is generated from lead/zinc production is estimated to exceed 6 million tonnes. When the waste slag is piled up, the heavy metal in the slag migrates and causes serious pollution in the surrounding water. At present, heavy metal pollution in China has become an increasingly serious and prominent environmental problem that causes serious damage to the health of citizens [36]. Magnesium slag and fly ash-based geopolymers are used to solidify/stabilize heavy metals in acidic residue, and this can prevent the heavy metal pollutants in waste from entering the environment again and causing secondary pollution. Ultimately, this achieves the purpose of treating waste using waste and the efficient utilization and savings of resources.

5.4.2 Morphological Analysis of Heavy Metals Pb and Cd in Acidic Residue

The BCR three-step extraction method was used to perform leaching and extraction experiments on acidic residue samples. It was found that the morphology distribution of Pb in the acidic residue was in the order of oxidizable state > reducible state > residue state > extractable state by acid. Among these states, the oxidizable state accounts for 69.02% of the total Pb. The state distribution of Cd in the acidic residue is in the order of extractable state by acid > oxidizable state > reducible state > residue state; among these, the content of the extractable state by acid accounts for 80.89% of the total Cd. The doping of magnesium slag promotes the conversion of the heavy metals Pb, Cd, Cu, and Zn in the waste slag from an unsteady state to a stable state, and this achieves the purpose of solidifying/stabilizing the heavy metals Pb, Cd, Cu, and Zn [34].

5.4.3 Experiments and Research Method

Main research methods: (1) Chemical analysis methods and X-ray fluorescence spectrometry (XRF) are used to determine the main chemical compositions and the percent of each compound in industrial solid wastes such as magnesium slag, fly ash, and acidic residue; X-ray diffraction (XRD) is used to determine the main phases of the solid wastes. (2) The morphology distribution of Pb, Cu, and Cd in acidic residue is analyzed using the BCR long-range order extraction method; magnesium slag is used to solidify/stabilize heavy metals in acidic residue, and the morphology distribution of heavy metals in the solidified/stabilized waste residue is analyzed. Toxicity leaching experiment are performed on the samples, and inductively coupled plasma emission spectrometry (ICP-7000) is used to determine the contents of heavy metals in the leaching solution. XRD, scanning electron microscopy with Energy dispersive X-ray spectroscopy (SEM/EDX) and Fourier transform infrared spectroscopy

(FTIR) are used to study the phase composition and microstructure of the waste residue before and after the toxicity leaching experiment. (3) Fly ash and alkali activator are used to prepare fly ash-based geopolymer. The prepared geopolymer that has good mechanical properties is used to solidify/stabilize the heavy metals Pb and Cd. The mechanical properties of the fly ash-based geopolymer are determined, and a toxicity leaching test is carried out to determine the amounts of Pb and Cd in the leached solution.

The experimental research is divided into four parts: (1) solidification/stabilization of heavy metals experiment using magnesium slag and acidic residue at high temperature; (2) solidification/stabilization of heavy metal experiment using magnesium slag and acidic residue at room temperature; (3) preparation of fly ash-based geopolymers; (4) solidification/stabilization of heavy metals using the fly ash-based geopolymers. The process routes are shown in Figs. 5.8, 5.9, 5.10, 5.11.

The contents of heavy metal elements in magnesium slag and acidic residue are shown in Table 5.4. As seen in Table 5.4, the amounts of harmful heavy metal elements in magnesium slag are relatively low. Thus, the environmental hazards caused by heavy metals in magnesium slag can be ignored. Also, the amounts of the heavy metals Pb, Zn, and Cd in acidic residue are relatively high.

$Pb(NO_3)_2$ and $Cd(NO_3)_2 \cdot 4H_2O$ are dissolved in distilled water to prepare solutions with different concentrations of Pb and Cd. The solutions are added to the original acidic residue; then, the mixture is stirred, dried, and ground to obtain acidic residue

Fig. 5.8 Process flowchart of solidifying/stabilizing heavy metals at high temperature

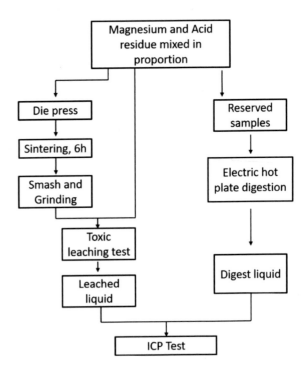

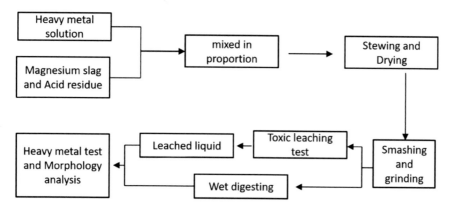

Fig. 5.9 Process flowchart of solidifying/stabilizing heavy metals at room temperature

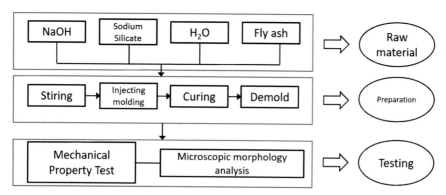

Fig. 5.10 Preparation of fly ash-based geopolymer

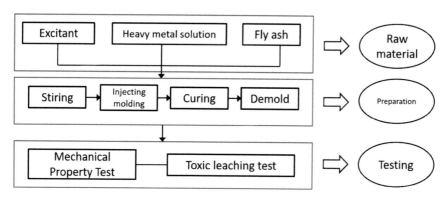

Fig. 5.11 Process flowchart of solidifying/stabilizing heavy metals using fly ash-based geopolymer

Table 5.4 Contents of heavy metal elements in magnesium slag and acidic residue (%)

Element	Pb	As	Cd	Cr	Cu	Zn	Co	Ni
Magnesium slag	0.002	ND	ND	0.010	0.002	ND	0.0004	0.010
Acidic residue	0.995	0.17	0.315	0.006	0.023	1.380	ND	0.0006

(ND: not detected)

samples with different concentrations of Pb and Cd. The sulfuric acid and nitric acid method (HJ/T299-2007) is used to perform a toxicity leaching experiment. For the leaching solutions obtained in the toxicity leaching experiments, the Pb concentrations in all of the samples (including the original acidic residue without added Pb), exceed the concentration limit of Pb $\leq$ 5 mg/L in GB5085.3-2007. The Cd concentrations of all of the samples with added Cd exceed the limit of Cd $\leq$ 1 mg/L in GB5085.3-2007. It is observed that the amounts of the heavy metals Pb and Cd in the acidic residue discharged from lead-zinc plants are relatively high. Although the enriched amount does not have a recycling value, the leaching toxicity far exceeds the national standard, and thus, it needs to be solidified/stabilized before it is discharged.

5.4.4 Solidification/Stabilization of Heavy Metals Using Mg Slag and Acidic Residue at High Temperature

To stabilize the heavy metals in the acidic residue and to reduce their environmental pollution, magnesium slag was mixed with acidic residue to achieve the fixation/stabilization of heavy metals at high temperature according to the characteristics of magnesium slag.

Acidic residue (acidic residue + heavy metal) samples with heavy metal content exceeding the limits in the standard were prepared. The samples were mixed with magnesium slag at a ratio of 40:60%. The mixture was pressed into blocks using a hydraulic press, and they were sintered in a box-type resistance furnace at a temperature of 1200 °C. The temperature was maintained for 6 h; then, after the samples were cooled in the furnace, they were removed, and the toxicity leaching experiment was carried out. The experimental results indicate that when the amounts of Cd, Cu, and Pb reach 1.070, 2.471, and 0.38%, respectively, in the samples, the concentrations of Cd, Cu, and Pb in the leaching solution meets the critical limit in the GB5085.3-2007 standard. This shows that the Cd, Cu, and Pb in the samples can exist stably and are not easily leached after the samples were doped with 60% magnesium slag and sintered at 1200 °C for 6 h.

XRD phase analysis shows that after sintering, a part of γ-C$_2$S phase in the sample transformed into the β-C$_2$S phase, and this increased the amount of β-C$_2$S and decreased the amount of γ-C$_2$S. This is because γ-C$_2$S is the stable phase of Ca$_2$SiO$_4$ at normal temperature, and β-C$_2$S is the high-temperature stable phase of

Ca_2SiO_4. After high temperature treatment, the γ-C_2S in the sample converted into β-C_2S. A higher amount of β-C_2S crystalline phases is conducive to the solidification/stabilization of heavy metals, such that heavy metals are stabilized in slag and are not easily leached. Compared with the nonsintered sample with the same composition, the leaching solution of the sintered samples had lower observed concentrations of heavy metals in leaching solution.

5.4.5 Solidification/Stabilization of Heavy Metals Using Mg Slag and Acidic Residue at Room Temperature

A certain quality of heavy metal salts containing Pb, Cu, and Cd was dissolved in distilled water to obtain heavy metal solutions with different concentrations. The solutions were poured into a mixture of magnesium slag and acidic residue to ensure that heavy metals were present as free ions in the slags. After standing for 5 min, the samples were dried, crushed, and ground. The treated samples were subjected to toxicity leaching experiment using the sulfuric acid and nitric acid method (HJ/T299-2007).

The main component of magnesium slag is C_2S, and its structure is a metastable high-temperature structure. Magnesium slag has a high hydration activity and can form calcium silicate hydrate gel (C-S-H gel) after hydration. Consequently, the heavy metals in the slag can be effectively solidified/stabilized in the slag and cannot be easily leached when the acidic residue is doped with magnesium slag.

The Pb in the original acidic residue was unstable and is easily leached. For the original acidic residue without doped with magnesium slag, when the Pb content in the original acidic residue reached 1.810%, the Pb concentration of the leaching solution exceeded the limit determined by GB5085.3-2007. For the same residue sample but was doped with 80% magnesium slag, the Pb concentration met the standard. Even, when the original Pb content increased to 2.650% in the sample doped with 80% magnesium slag, the Pb concentration of the leaching solution still met the standard. These results indicated that doping with 80% magnesium slag, Pb in the slag material is stable and not easily leached.

When the acidic residue is mixed with 10 and 20% magnesium slag, both Cd and Cu are stable in the slag and are not easily leached. The leaching rate is defined as the ratio between the heavy metal content of the leaching solution of treated samples and that of the original slag. When the leaching rate is smaller, the fixing/stabilizing effect is better. Experimental results indicate that when the doping level of magnesium slag is 10 and 20%, all of the Pb, Cu, and Cd concentrations of the leaching solution of acidic residue meet the GB5085.3-2007 standard.

References

1. Liu Y (2006) Preparation and study on slag glass-ceramics. Dissertation, Hunan University
2. Su L (2018) Study on properties of glass-ceramic made from magnesium reduction slag. North Minzu University, Thesis
3. Nan X (2006) Preparation of glass-ceramic. Dissertation, Lanzhou University of technology
4. Yoon S, Lee J et al (2013) Characterization of wollastonite glass-ceramics made from waste glass and coal fly ash. J Mater Sci Technol 29(2):149–153
5. Chen F, Zhao E et al (2007) The research development and application of $CaO-Al_2O_3-SiO_2$ glass for decoration. Ceramics 2(20):3–4
6. You X (2014) Review on the preparation of glass-ceramics from fly ash. Bullet Chinese Ceramic Soc 33(11):2902–2907
7. Chen G, Liu X et al (2002) Slag glass-ceramics: fabrications and propects. ceramics 4(16):2–3
8. Chen H (1988) Study on slag based glass ceramics. Glass and Enamel 16(2):1–7
9. Li J, Qian W et al (1992) Machinable glass-ceramics made from industrial slags. J Inorgan Mater 7(2):3–4
10. Chen G (1993) Microcrystallization of $CaO-Al_2O_3-SiO_2$ Syst. Glass 6:1–6
11. Zhao B (2010) Experimental study on glass-ceramic preparation from sewage sludge ash by microwave melting method. Dissertation, Harbin Institute of Technology
12. Ma X (2011) Classification and application of polarization glass. World of Building Mater 32(3):12–15
13. Yao Q (2005) The process and property investigation of steel slag glass-ceramics. Dissertation, Nanjing university of technology
14. Xiao HN, Deng C et al (2001) Effect of processing conditions on the microstructure of glass-ceramics prepared from iron and steel slag. J Human Univer Natural Sci 28(1):32–35
15. Zeng L, Hu D et al (2008) The novel techniques and development of preparation of porous ceramics. China Ceramics 44(7):7–11
16. Huang X, Ma X et al (2015) Current situation of preparation and application of porous ceramic materials. China Ceramics 9:5–8
17. Ju Y, Song S et al (2007) The preparation, applications and research development of the porous ceramics. Bullet Chinese Ceramic Soc 26(5):969–976
18. Li X, Zhang S et al (2011) Review on the recycle of magnesium slag wastes. Concrete 8:97–101
19. Lei R, Fu D et al (2013) Research progress of fly ash comprehensive utilization. Clean Coal Technol 3:106–109
20. Wang H, Tong J et al (2007) Resourcification utilization routes for carbide slag. Chem Prod Technol 1:47–53
21. Hai W, Han F et al (2018) Influence of ratio of raw materials on the properties and morphology of industry solid wastes porous ceramics. Bullet China Ceramic Soc 37(12):3776–3780
22. Xiao L, Luo F et al (2009) The analysis on mechanism of using magnesium slag to prepare the cementitious material. J Jilin Jianzhu Univer 26(5):1–5
23. Han F, Zhou S et al (2013). Cementing material preparation from magnesium slag. In: Proceeding of annual conference of chinese society for environmental sciences, pp 5554–5557
24. Zhang H, Li H et al (2014) Preparation of Belite-sulphoaluminate cement using phospho-gypsum. Bullet China Ceramic Soc 33(6):1567–1571
25. Arjunan P, Michael R et al (1999) Sulfoalu minate-belite cement from low-calcium fly ash and sulfur-rich and other industri-al by-products. Cement Concrete Res 29:1305–1311
26. Wan X, Zhang M et al (2010) Study on treatment of metallurgical solid wastes by sulphoaluminate cement. Environ Eng 01:73–76
27. Xu G (2009) Research of utilizing coal gangue in Shi zuishan district to produce series of sulphoaluminate cements. Dissertation, Chengdu University of Technology
28. Zhao S, Han F et al (2017) Preparation of composite slag sulphoaluminate cement clinker from electrolytic manganese-magnesium. Bullet China Ceramic Soc 36(05):1766–1772 + 1776
29. Yuan Y, Ye Z et al (2012) Influence of iron phase composition on barium calcium sulphoaluminate cement. J Univer Jinan (science and technology) 26(2):128–131

30. Feng X, Zhu Y et al (1984) Investigation on strength development of C_4AF and a new type of cement with high early strength and high iron content. J Chinese Ceramic Soc 12(1):32–47
31. Zhou S (2015). Solidification/stabilization of heavy metals in industrial solid waste. Dissertation, Jiangsu university of technology
32. Chen Y, Han F et al (2015) Solidification/stabilization of heavy mental Cu and Cd in waste acid residue by magnesium slag. Inorgan Chem Indus 47(7):48–51
33. Du Y, Jin F et al (2011) Review of stabilization/solidification technique for remediation of heavy metals contaminated lands. Rock Soil Mech 32(1):116–124
34. Chen Y (2016) Research on the solidification and stabilization of heavy mental by magnesium slag and geopolymer based on fly-ash. Dissertation, North Minzu University
35. Chen Y, Han F et al (2016) Solidification and stabilization of heavy mental Pb in waste acid residue by magnesium slag. Chinese J Environ Eng 10(06):3229–3234
36. Zhang J, Wei J et al (2014) Legislation outline on heavy metal pollution prevention. Environ Sustain Develop 1:60–62

Printed in the United States
by Baker & Taylor Publisher Services